American Institute of Chemical Engineers

Conference of the Computing and Systems Technology Division 2007

Held at the 2007 AIChE Spring National Meeting

April 22-27, 2007
Houston, Texas, USA

Printed from e-media with permission by:

Curran Associates, Inc.
57 Morehouse Lane
Red Hook, NY 12571
www.proceedings.com

ISBN: 978-1-60423-363-6

ISBN: 978-1-60423-363-6

American Institute of Chemical Engineers
Proceedings
Three Park Avenue
New York, NY 10016-5991
Phone: 212-591-8100

www.aiche.org

American Institute of Chemical Engineers

Conference of the Computing and Systems Technology Division
2007

TABLE OF CONTENTS

Session 11 – Process Design Software and Case Studies

Session 27 – Advances in Optimization for Process Operations, Planning and Scheduling

Session 57 – Process and Control System Monitoring

Session 76 – Process Control and Monitoring Applications

Combined Pre-Concentrator/Recovery Column Design for Ethanol Dehydration Process

Saiful Arifin and **I-Lung Chien**. Department of Chemical Engineering, National Taiwan University of Science and Technology, No. 43, Keelung Road, Sec. 4, Taipei, 106, Taiwan

Ethanol dehydration process is important in producing high-purity ethanol from renewable biomass resources. In a diluted stream from fermentation after filtration or centrifuge to remove heavy phase materials, the ethanol content is typically no more than 10 mol%. Taking a feed stream to the ethanol dehydration system with 5 mol% of ethanol and 95 mol% of water as an example, this paper will investigate the optimal design flowsheet to produce high-purity ethanol via heterogeneous azeotropic distillation. A classical example of ethanol dehydration system is to use benzene as entrainer to aid the separation. However, because benzene is a known carcinogen, this work will demonstrate the design concept using cyclohexae or n-hexane as entrainer.

An obvious design for the overall process is to include a pre-concentrator column in the upstream of the heterogeneous azeotropic column to concentrate the fresh feed composition to a point near the azeotropic composition of ethanol and water. After the pre-concentrator column, with the help of the entrainer entering into the heterogeneous azeotropic column, the top vapor of this second column will be close to the lowest boiling point of the ternary system (ternary azeotrope with b.p of 62.39 oC using cyclohexane as entrainer). The bottom product of this heterogeneous azeotropic column is pure ethanol. The top vapor after condensation can naturally separated into two liquid phases with the organic phase containing mostly entrainer which can be recycled back to the heterogeneous azeotropic column. The aqueous phase still contains significant amount of ethanol will enter into recovery column. The bottom stream of the recovery column is pure water while the distillate can be designed to recycle back to heterogeneous azeotropic column. This conventional overall process will contain three distillation columns (pre-concentrator column, heterogeneous azeotropic column, and a recovery column).

An alternative simpler process design is proposed in this paper with a column (C-2 in Figure 1 below) acting as both the pre-concentrator column and the recovery column. This design flowsheet can be seen in Figure 1 with material balance lines shown in Figure 2 using cyclohexane as entrainer. The Total Annual Cost (TAC) will be used as the objective function to be minimized for the above two alternative process design flowsheets (conventional three-column system and proposed two-column system). A comparison of the TACs will be made for the optimized two systems. Because feed composition of the waste IPA stream can vary widely, the overall control strategy for these two alternative design are critically important in rejecting this feed composition disturbance. The operability of these two alternative process systems under proper overall control strategy will also be compared. Aspen Plus and Aspen Dynamics will be used in the above design and control studies.

Keywords: Ethanol Dehydration, Heterogeneous Azeotropic Distillation, Pre-concentrator column, Recovery Column, Design and Control.

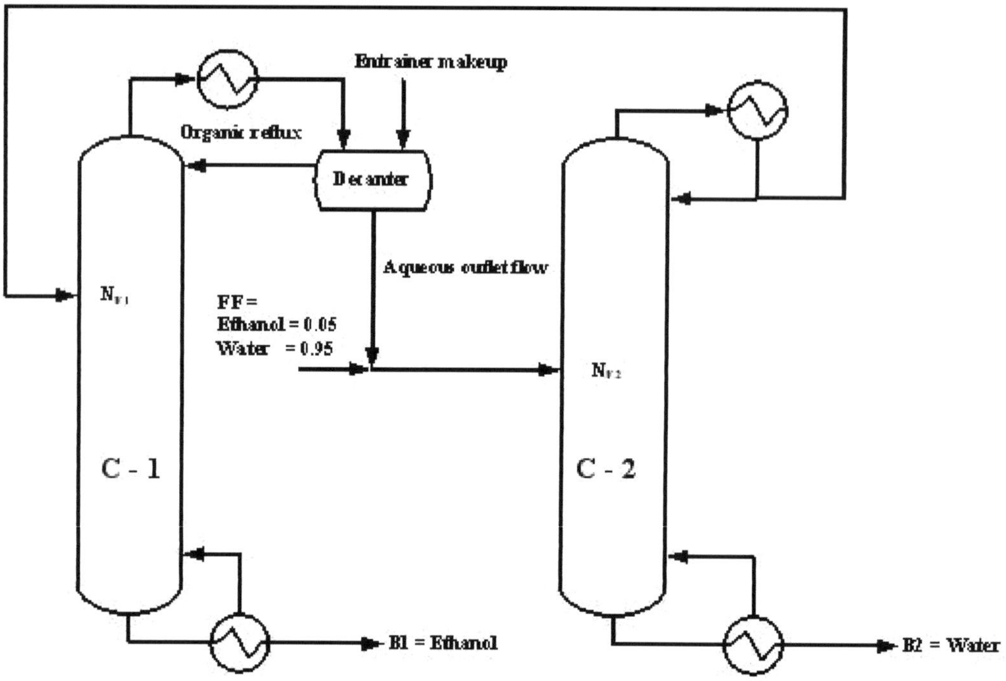

Figure1. Proposed process flowsheet of the two-column system.

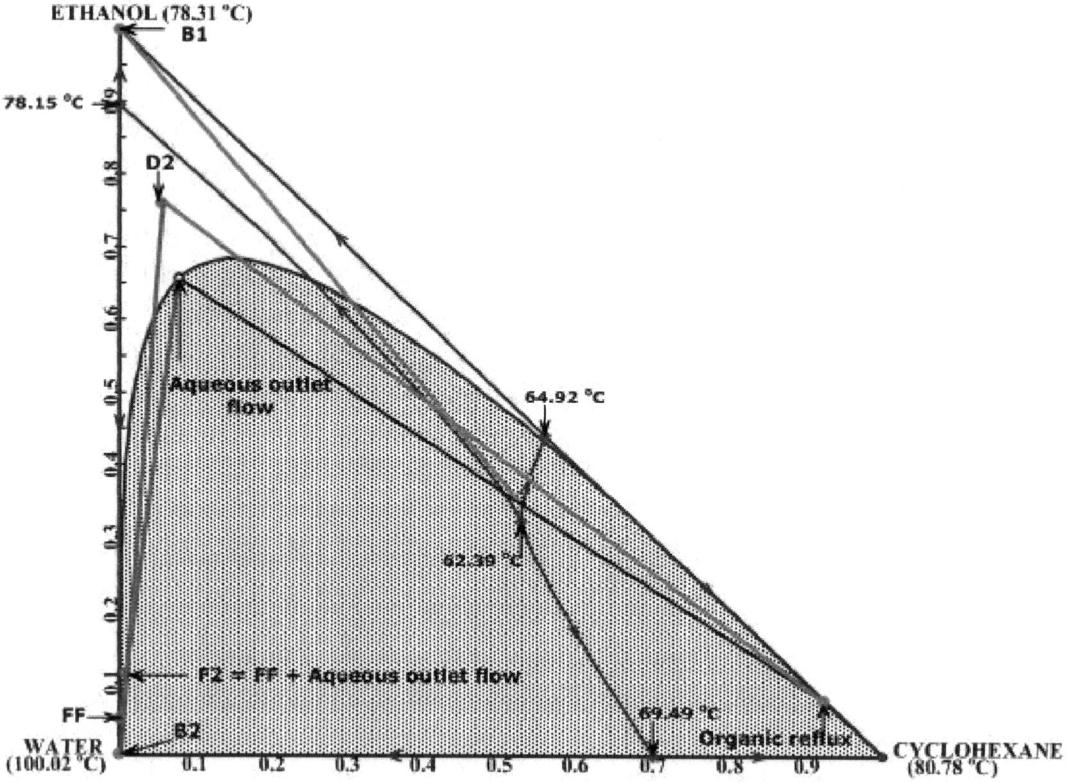

Figure 2. Conceptual Design of the proposed two-column system.

MODELING FUEL DESULFURIZATION BY ADSORPTION VIA A PROBABILITY METHOD

Kimberly Drake, John Heinzel, and Ian Peek, Naval Sea Systems Command – Philadelphia, PA
Donald Hoffman, Office of Naval Research, Arlington, VA

Introduction

The Office of Naval Research (ONR) and the Naval Sea Systems Command (NAVSEA) are jointly sponsoring an Advanced Technology Development program to demonstrate a Ship Service Fuel Cell (SSFC) electrical generator in the marine environment. This technology is managed under the ONR Advanced Electrical Capabilities "ACES" Future Naval Capability (FNC) and will contribute to deployment of the US Navy's first "All-Electric Ship".

There are many benefits of fuel cell generators to ship operations, including reduced fuel consumption at full and partial loads, smaller intake and exhaust ducting, and reduced thermal and acoustic signatures (Nickens, et al 2006). Under this program a 625 kW Molten Carbonate fuel cell generator, and a 500 kW Fuel Processor, designed for Proton Exchange Membrane (PEM) fuel cell applications, were demonstrated. Both systems are focused on demonstration of a fuel cell based ship service generator using NATO-F76 and JP-5 logistics fuel. A lesson learned from the systems is that certain subsystems should be improved in an effort to significantly increase power density while also improving system lifetime and overall robustness (Heinzel et al 2006).

To this end, ONR is sponsoring advanced research in a variety of component areas with the goal of obtaining a three to four time improvement in overall system power density. One of the most critical areas being explored is the ability to remove sulfur from F-76 and JP-5 Navy logistics fuels in the liquid phase at ambient temperature and pressure with high regeneration capability over high numbers of cycles (Nickens, et al 2006). Removal of refractory sulfur compounds is a challenge due to the nature of the molecules themselves, as well as concentrations, as high as 10000 ppm (Song and Ma 2003).

The desulfurization process is essentially an adsorptive process. A simple example of an application of adsorption is the use of charcoal filters in the process filtering water. Water passes through the torturous bed of charcoal in the filter. Impurities remained attached to the charcoal as the water passes. Desulfurization of fuel proceeds analogously.

Previous systems used hydrodesulfurization (HDS) and gas-phase desulfurization systems to prevent sulfur in the fuel from poisoning the fuel cell system. Hydrodesulfurization is a catalytic treatment process which utilizes hydrogen to convert sulfur compounds present in liquid fuel to slightly different compounds plus hydrogen sulfide gas. The process requires relatively high temperature and pressure conditions (Miller et al 2006). Hydrogen sulfide is then separated and removed from the process stream, either through gas-phase desulfurization or conversion to elemental sulfur.

Gas-phase desulfurization is a regenerable sorption process where typically a metallic oxide sorbent is contacted with gas that includes hydrogen sulfide to produce a metal sulfide, essentially exchanging oxygen for sulfur. The process occurs at elevated temperature and pressure, requiring that a high-sulfur fuel be reformed or broken down (in an HDS reaction) into gaseous products prior to sulfur removal (Heinzel et al 2006). This effectively limits the reformation technologies that can be utilized, effecting overall efficiency and size when used as the primary means of desulfurization or increasing required space when utilized as part of the HDS process.

Unfortunately, hydrodesulfurization requires additional pure hydrogen, as well as relatively severe reaction conditions. Gas-phase removal of sulfur as hydrogen sulfide (H_2S) through sorbent beds proved to be ineffective due to the large size beds required to handle hot reforemate gases at pressure (Heinzel et al 2006). Additionally, in order to provide an adequate service life, the gas-phase sulfur removal sorbent beds must be able to be regenerated or cleaned. Under such conditions, the valves and actuators necessary to switch operating beds are extremely large, severely reducing power density.

Overall, hydrodesulfurization and gas-phase desulfurization systems require extreme temperature conditions and add unacceptable weight to the system, making both processes unsuitable for shipboard use (Nickens, et al 2006). The ability to remove sulfur from fuel in the liquid phase at ambient temperature and pressure would allow systems to become far more compact due to the elimination of pressure vessels, large valve and actuator assemblies and high inventory reactors, which contain large amounts of vaporized fuel and other combustibles.

Design of sorbent beds to maximize mass contact and heterogeneous contacting efficiencies is critical to the realization of beds with high regenerative performance and maximum capacity. Central to this is the need to understand molecular flow and interaction with sorbent particle surfaces. The combination of understanding of theoretical sorbent capacity and probabilistic estimation of bed performance as sulfur loading increases will help to estimate the design parameters that will produce a bed with optimized capacity while minimizing pressure drop, thus reducing pumping losses and inefficiencies in operation.

Probabilistic Model Of Desulfurization

Introduction to the Probability Problem

Because the desulfurization of fuel through the process of adsorption is fundamentally important to fuel cell technology, one current project is the development of a model of the probability that sulfur will be adsorbed as fuel passes through the adsorbent bed. As discussed above, there are several processes used for the desulfurization of fuel. Because of the variations in these processes, the first objective in developing the probability model is to keep the model as general as possible and to develop as a series of working models that can be modified or tailored to a particular process. The benefit of taking this approach is that the model can be further refined for different processes from any one of the intermediate stages.

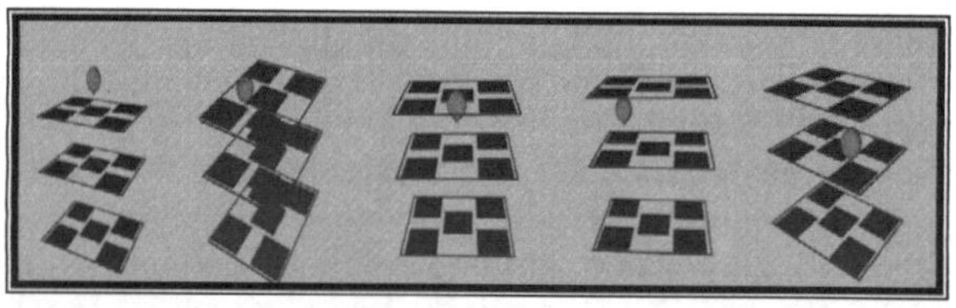

Figure 1: Graphical representation of the basic geometry for the initial model

Developing the Model Based on Geometry

The desulfurization process will be viewed geometrically first. The sorbent bed can be thought of as a cylinder packed with particles. The fuel passes around the particles losing its sulfur compounds to adsorption on the particle surfaces along the way. The process has 3 main geometric elements. The first element is a molecule of fuel, initially modeled as a sphere. While it's true that fuel molecules can have highly irregular shapes, it's not unreasonable to model them as a sphere as a first order approximation. The second element is the sulfur compound attached to the fuel, modeled as a cube attached to the sphere. Given the shape of the molecule and its multitude of potential orientations, the position of the sulfur on the molecule is taken to be unknown. Finally, the last element of the initial geometry will be a differential slice representing the adsorbate. It is assumed that on the surface of the slices there are sites to which the sulfur may adsorb. If the sulfur does not attach to the site, then it is assumed that it passes through the slice. Figure 1 is a graphical representation of the process of the molecule of fuel passing through a sequence of slices.

A rough estimate of the probability that the sulfur is removed from the fuel by passing through the slice can be found by calculating the probability that the sulfur on the molecule is oriented toward the slice and that molecule contacts one of the sites. Then, calculating the probability of the sulfur being adsorbed to a location on the slice is:

$$P = P(\textit{sulfur is properly oriented}) * P(\textit{molecule hits the site on slice}).$$

Assume that the positioning of the sulfur is random and that the position on the slice where the molecule touches is also random. Obviously, this method does not take into account many things. It does not consider the properties of any of the chemicals, their attraction to each other, what 'oriented toward the slice' means, or how close is close enough in the orientation for sorption to occur. However, the purpose of this initial development is to build a working model characterizing the geometry of the adsorption process in a packed bed. Thus, while these assumptions are non-trivial and neglect much of the physics and chemistry, they do serve a purpose.

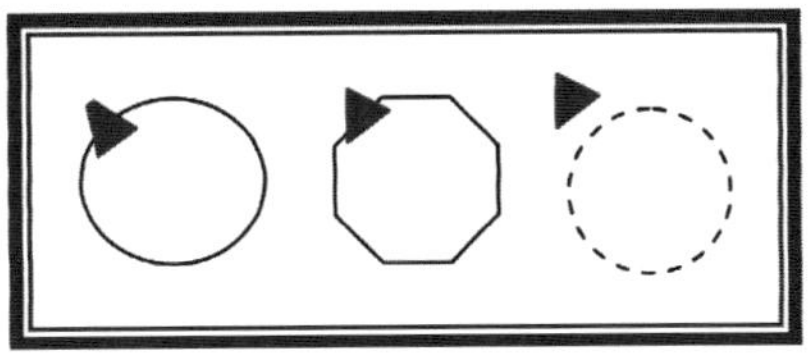

Figure 2: Illustration of discretization

In order to find the probability that the sulfur is oriented toward the slice and the probability that the molecule touches the site on the slice, the geometry of the slice and the sphere will be discretized. The model will be developed in sections, focusing first on the sphere and sulfur and then the slice. Once the probability of the sulfur being adsorbed on an open, active site as it passes through a slice has been calculated, it can be used to calculate the probability of the sulfur being adsorbed after passing through multiple slices and thus, can be used to model the entire adsorption process.

Discretization

In this section, the basics of discretizing the geometry of an object and its relationship to probability will be discussed by applying it to a small example. Then in subsequent sections, these ideas will be applied to the sphere and the slice.

Consider a circle or ring with a spot corresponding to exposed sulfur on the outer edge as seen in Figure 2. Further, suppose the objective is to calculate the probability that if one were to randomly touch some part of the circle, one would touch the exposed sulfur. In order to calculate this probability, a technique first used by the Greek mathematician, Archimedes, is employed. Observe that one way to model a circle mathematically is to consider a sequence of regular polygons. As is seen in Figure 3, as number of sides, goes to infinity, the n-gon converges to a circle.

This idea will be valuable because it allows for the reduction of the continuous figure of the circle to an object with a finite number of sides of equal length. Because the sides are of uniform length, each of the sides are equally likely when considering a probability. Using this idea of the polygons, the probability of touching upon the sulfur can be estimated to any degree of **accuracy** required.

As is seen in Figure 2, the probability that the sulfur molecule is on any one side of the n-gon is $1/n$ while the attachment point of the sulfur to an adsorbent surface is smaller than the side of the n-gon. Once the attachment point becomes larger than the side of the n-gon, the

Figure 3: Illustration of converging discretization.

probability of touching the sulfur is no longer $1/n$. Consider the rightmost polygon in Figure 4. The sulfur compound is actually on multiple sides of the n-gon. However, when the length of the side of n-gon becomes smaller than the size of the attachment point, the probability of touching upon the sulfur will converge.

Figure 4 illustrates how n=8 is a sufficiently fine discretization of the polygon in order to determine the probability of touching upon the position of the sulfur compound. The sulfur compound is touching one side of the octagon, in the figure. If a 24-gon were used rather than an octagon, the sulfur touches three of the sides. The probability calculated using the octagon would be 1/8; using the 24-gon, it would be 3/24=1/8. As is illustrated in the figures, using a discretization of the circle reduces the very difficult problem of estimating the probability of sulfur being on some random part of the circle down to a very easy calculation.

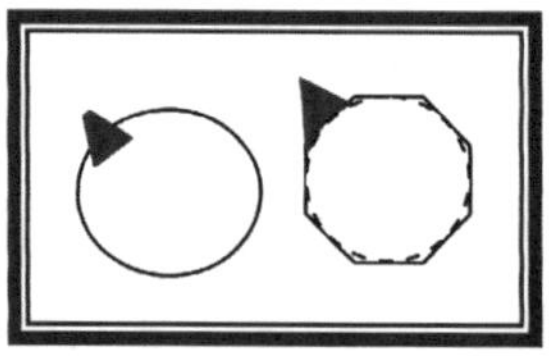

Figure 4: Illustration of refined discretization

The Molecule

Having discussed the basic ideas behind discretizing and the benefits of doing so, the discussion turns to discretizing the sphere. First, in moving from the discretized circle to the discretized sphere, one can not simply use the analogous 3-d case and estimate the sphere with polyhedrons. In order to ensure that any side of the discretized sphere is equally likely to any other side, the sides must be regular[1]. The 5 platonic solids are the only regular polyhedron. However, we can discretize the sphere in such a way that once the number of sides becomes sufficiently large, any variation in the size of the polygons composing sides of the estimation of the sphere will be so small as to be negligible.

Distributing Points on the Sphere

Problem: The discretization of the sphere needs to be a nearly regular partition which will become increasingly accurate as a function of an integer, *K*. That is, the partition must converge to the sphere as *K* goes to infinity.

***Solution*:** Distribute a number of points across the sphere at a fixed distance and then use those points as the centers of tangent squares.

Distributing points across the sphere evenly is a complicated problem. This is an active area of research and there are many algorithms for distributing points given a particular criterion for the distribution. For the purpose of this work, the optimal distribution of points is the one which ensures that each point is no closer than some fixed distance along the surface of the sphere. Unfortunately, this is an unsolved problem. However, the solution can be estimated to a sufficient degree of accuracy for the purposes of this model.

The algorithm used is a modified version of the algorithm found at (Ruskin 2006). An important addition in the algorithm is that the choice of the number of points generated is based on the size of the square. Later, the size of the square will be based upon the size of the sulfur.

Algorithm

1. Choose *L*, the side length of the desired square.
2. Choose or set *R*, the radius of the sphere.
3. Calculate the central angle
4. Calculate the arc length associated with L and R. (arc length=Radius * central angle)
5. Calculate *K*, the number of "arcs" on the sphere
6. Calculate the revised central angle
7. Calculate the revised arc length (i.e. one that divides the sphere)
8. Distribute *K*+1 points on the meridian (North Pole to South Pole) of the sphere.
9. Envision those points being swirled around the sphere. Each point marks a circle in the horizontal.
10. For each circle,
    a. Figure the height of the circle on the z axis.
    b. Calculate the radius, r, of that circle
    c. Calculate the circumference of the circle

---

[1] **Regular** - all angles are equal and all sides are the same length. Regular polygons are both equiangular and equilateral.

d. Using the fact that Arc Length=(Angle Subtended by the Arc)*(Radius)= φ/r, calculate the angle with revised arc length $\frac{R\pi}{K}$, and radius, r.

e. Calculate the number of arcs on this circle by calculating Circumference/arc length.

Worthy of particular note is the initial choice of arc length, *L*. The algorithm allows the user to choose any value of *L* and this is likely to generate error as it's probable that the given value of *L* will not divide the sphere evenly. For example, if the user chose a value *L*= .1 and the radius of the sphere was *R*=2, then the half circumference of the sphere (*C/2*=6.28318530717958645) divided by the given value of *L* would be 62.83. This would mean that there were 62.83 arcs when the sphere is divided. This translates to 62.83 circles in the second part of the algorithm. Clearly, it's necessary to have a whole number of circles.

Rather than having a relatively large 'patch' of error in the last row, arc length is recalculated in the initial stages of the algorithm in order to distribute it evenly among the squares. That is, the size of the squares is adjusted from where the user wanted it to be. As is seen in Table 1, once *L* gets small, the adjustment is negligible. One might argue that it's introducing error into the model by making this adjustment. While that is true, this is a better way of introducing error into the model than the alternative of having partial counts. By adjusting the size early in the process, the error shrinks with the size of the squares.

The results of this algorithm for several values of *L* and a sphere of radius 1 are shown in Table 1. As *L* decreases, the number of points increases incredibly rapidly. It's also clear that as *L* decreases, the difference generated by recalculating the arc length to be something which divides the sphere dramatically decreases. The difference in size between the surface area of a true sphere and the surface area of the estimated sphere also decreases dramatically after only a few values of *K*.

Note that there is likely to be a small amount of error at each horizontal circle approximately at the point of the original meridian for reasons similar to the one described above. The error is introduced into the problem when points are set on the horizontal circles (step 9 in the algorithm above). Each of those circles has a different circumference. It's unlikely that the revised arc length divides those circumferences evenly. To avoid this problem the algorithm rounds. That is, if the left over length on the circumference is more than half the needed distance, then another point is put on the circle. If it is smaller than the needed distance, we do not. There are other choices in dealing with this error. If the *floor* function were used, there would be less points and a less conservative estimate. If the ceiling function were used, there would be more points and a more conservative estimate. Since this error occurs only once at each horizontal circle, the count of our points is in error for some number of points between ±K.

Table 1 Results of distributing points on the sphere

Desired Arc Length	Calculated Arc Length	K	Number of Points	Sum of Square Areas	|Sphere Surface Area-Square Area|	|Estimated Square Area-Exact Square Area|
1.0	1.04719755119660	3	10	10.96622	1.60014e+000	9.6623e-002
0.1	0.10134169850290	31	1,220	12.52957	3.67999e-002	2.7014e-004
0.01	0.01000507214519	314	125,536	12.56633	3.26483e-005	1.0147e-007
0.001	0.00099987035442	3,142	12,569,664	12.56640	3.43941e-005	2.5927e-010
0.0001	0.00009999976616	31,416	1,256,642,920	12.56637	1.85708e-007	4.6769e-014

It may seem like having $\pm K$ points as an error is large, especially when the value of K gets large. However, it must be remembered that K is not the number of points on the sphere but used only to calculate the number of circles imposed on the sphere upon which the points will be distributed. The error that is generated for a large K is negligible. For example, in Table 1, note that for the choice of L=.0001, 1,256,642,920 points are generated. If the error in the count is maximized, 1,256,674,336 points might have generated. If the sulfur is of size L, then the error in the probability is approximately

$$\left| \frac{1}{1{,}256{,}642{,}920} - \frac{1}{1{,}256{,}674{,}336} \right| = 1.98\text{e-}014$$

due to this $\pm K$.

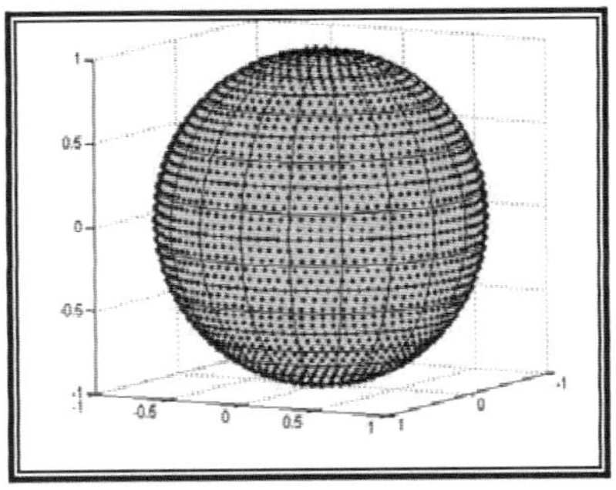

Figure 5: Illustration of points distributed on sphere.

This approximation of the sphere is ideal for the purposes of this model. Figure 5 illustrates a sphere with L=.001 and 3,142 points distributed across its surface. As seen in Table 1 the difference in area between squares of size L squares estimated is only 2.5927e-010 and the total area of the squares is 12.56640. The difference in the surface area between the square tiled sphere and a true sphere is only 1.85708e-007.

In order to use the discretization of the sphere to model the probability in question, a closed form of the count of how many squares there are for any given value of K is required. Based on the algorithm described above the number of squares for some value of K is given by

$$\text{Number of squares} = \sum_{i=1}^{K} round\left(2K\cos\left[-\frac{\pi}{2}+\frac{\pi}{K}(i-1)\right]\right)$$ where K is chosen so that $$K = \text{round}\left(\frac{\pi R}{sl}\right).$$

Having a closed form for the number of points on the sphere, the discussion now turns to counting the sites on the slice.

The Differential Slice

The next question addressed is where the molecule contacts the differential slice representing the adsorbate. Portions of the slice will "catch" the molecule whereas other portions of the slice will let the molecule pass through due to the fact that sorbent particles are packed randomly and have space between them due to their geometries. If it is assumed that the material in one slice of the overall packed bed is in the form of a thin, rectangular slice, a grid can be superimposed on the slice, discretizing it as seen in Figure 6. Using this count, the probability of the sulfur touching a portion of the slice where the molecule will be adsorbed to an open, active site can be calculated.

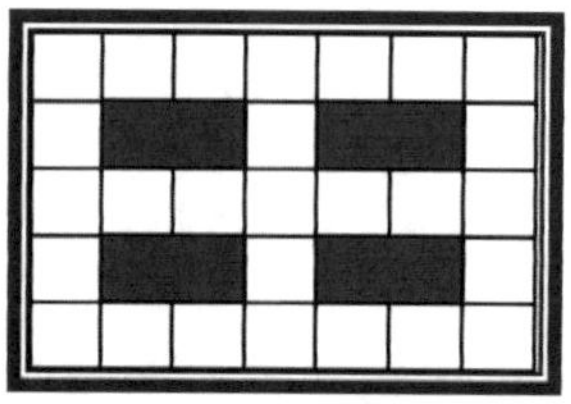

Figure 6: Illustration of discretiziation of the slice.

As seen in Figure 7, the probability that a molecule contacting the slice on one of the purple bins in easily calculated and is 8/35. By making the grid sufficiently fine, the probability can be calculated to any desired accuracy.

Alternative Discretizations

Note that for both the sphere and the slice, it is possible to get more accurate estimates of figures with irregular shapes by using a finer discretization. This alternative may be explored in more detail at a later date.

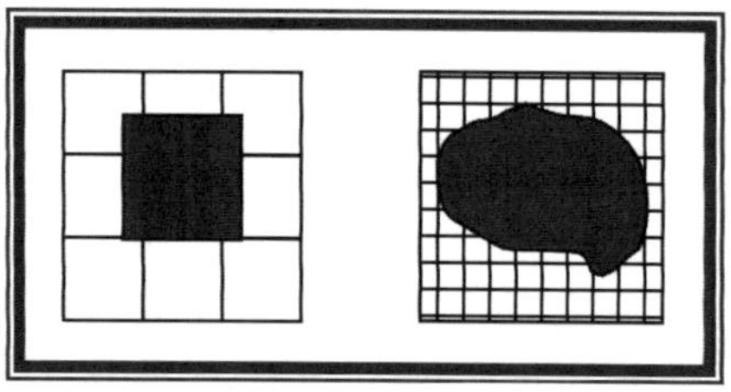

Figure 7: Illustration of refining discretization to accommodate irregular figures

Essentially, one would use a finer grid to more accurately capture the shape of the molecule and thereby capture a more accurate estimation of the probability. This is illustrated in the right grid in Figure 7. In the figure, the grid on the left illustrates how grid size is currently being chosen. Some fixed length, L, is given and grid size is chosen so that the squares in the grid have the closest size to L.

Bringing It All Together

In order to piece together the model, some notation and assumptions regarding the molecule and substrate need to be defined.

Some Assumptions and Notation

- Assume that all the sites and molecules, etc are of a uniform size for their kind.
- Let the slice that the molecule will pass through be represented by a rectangle of length and width *pl* and *pw* respectively. .
- Let the sites on the slice where the molecule can get 'caught' be considered squares on the slice and let them have length and width *sl*.
- Let the number of sites on the slice where the molecule can get caught be given by *ns*.
- Assume that the size of the site is bigger than the size of the sulfur molecule. Thus, $sl^2 > bl*bw$.
- Assume that the only way that the sulfur cube can connect to the site is that it is oriented perfectly toward the site. That is, if the sphere is turned a bit, then the sulfur cube will not be aligned to hit the site and adsorb.
- Assume that no outside forces are acting as attractors or repellers on the bulk molecules or sulfur atoms within specific compounds in the bulk. That is, there are no chemical or physical factors involved in the orientation of the sphere.

Using the discretized sphere and the discretized slice, a rough estimate of the probability that the sulfur is filtered from the fluid is found by calculating the probability that the sulfur molecule is oriented toward the slice and that molecule contacts a portion of the slice that will adsorb it. Then, to calculate the probability of the sulfur being caught on the slice, we calculate *P= P(sulfur is properly oriented)*P(molecule hits the slice)*.

Probability that the Sulfur Contacts the Slice

In the same way that the probability that the sphere is oriented toward the slice was calculated, so the probability that the sphere hits one of the sites when it touches the slice is calculated. According to the assumptions, the number of sites on the slice is *ns* and the slice is of size (*pl* x *pw*) where (*pl* x *pw*) is divisible by the size of the site, sl^2. Again, it is assumed that there are no attracting or repelling influences on the molecule so that contacting one of the sites is entirely random. Then, the probability of touching one of the sites is the ratio of the number of sites to the number of possible places to make contact. That is,

$$P_{Site}(ns, pl, pw, sl) = \frac{ns}{pl \times pw / sl^2}.$$

Probability That The Molecule Is Oriented Toward The Slice

The estimation of the probability that the molecule is oriented toward the slice is found by observing the following:

If a sphere constructed of squares is to touch a flat surface, then one of the squares will be oriented toward the surface and the other squares will be oriented elsewhere. The number of squares on the sphere is given by

$$\text{Num squares} = \sum_{i=1}^{K} \text{round}\left(2K\cos\left[-\frac{\pi}{2}+\frac{\pi}{K}(i-1)\right]\right)$$

where K is chosen so that $K = \text{round}\left(\frac{\pi R}{sl}\right)$. Then, the probability that the sphere is oriented toward the slice is the same as the probability of randomly choosing a square from the sphere. Thus, the probability is:

$$P_{Sphere}\left(K = \text{round}\left(\frac{\pi R}{sl}\right)\right) = \frac{1}{\sum_{i=1}^{K} \text{round}\left(2K\cos\left[-\frac{\pi}{2}+\frac{\pi}{K}(i-1)\right]\right)}$$

Initial Model

Having computed the probability that the molecule will contact an active site and the probability that the sphere will be oriented so that the sulfur adsorbs, the total probability can now be composed. Because these things are independent, it can simply be calculated as

$$P_{Total}(K,ns,pl,pw,sl) = P_{Sphere}(K)\,P_{Site}(ns,pl,pw,sl) = \frac{1}{\sum_{i=1}^{K} \text{round}\left(2K\cos\left[-\frac{\pi}{2}+\frac{\pi}{K}(i-1)\right]\right)} \cdot \frac{ns}{pl \times pw / sl^2}$$

Note that the probability of the sulfur *NOT* being caught by the sorbent is $1 - P_{Total}$.

Multiple Slices—The Whole Filter

As a next step, consider the adsorbent bed as a stack of these adsorbing slices. In order to calculate the probability that the sulfur will remain on one of the slices, remember that the probability of the sulfur adsorbing to one of the slices is 1-*P*(sulfur adsorbed on none of the slices). The probability that the sulfur is not captured on the first slice is $(1 - P_{Total})$. The probability that the sulfur is not adsorbed on the first slice or the second slice is $(1 - P_{Total})^2$. It's worth noting here that $(1 - P_{Total}) < 1$ and this value to any power will also be less than one. Using the reasoning above, if the adsorbent bed is composed of *N* slices, then the probability that the sulfur is not adsorbed on any of the slices is $(1-P_{Total})^N$. Thus, since

P(sulfur adsorbed on none of the slices)= $(1-P_{Total})^N$,

then

$$P(\text{adsorbed}) = 1 - P(\text{ sulfur gets caught on none of the plates})$$

$$= 1 - (1 - P_{Total})^N$$

$$= 1 - \left(1 - \frac{1}{\sum_{i=1}^{K} \text{round}\left(2K \cos\left[-\frac{\pi}{2} + \frac{\pi}{K}(i-1) \right] \right)} \frac{ns}{pl \times pw / sl^2} \right)^N .$$

Conclusion

Using some basic geometry, a model has been developed estimating the probability that a molecule containing sulfur is removed from a bulk fuel of hydrocarbon molecules as it passes through a single slice of adsorption material and through multiple slices. While the model is coarse, it is sufficiently general as to be easily modified to capture the specific properties of the chemicals and compounds used but still capture the essential geometry of the system.

In future generations of this model, time, temperature and the nature of the sites in the slices will be taken into consideration.

References

1. Nickens, M. Cervi, S. Abens, D. Hoffman. "US Navy Ship Service Fuel Cell Program". Fuel Cell Seminar 2006, Honolulu, Hawaii, 2006.
2. T. Erber, G. M. Hockney, "Equilibrium configurations of n equal charges on a sphere", J Phys A: Math, 1991.
3. R. H. Hardin, N. J. A. Sloane and W. D. Smith http://www.research.att.com/~njas/electrons/
4. John Heinzel, Mark Cervi, Donald Hoffman and Anthony Nickens. "Advances in Logistic Fuels Desulfurization". 232 National Meeting of the American Chemical Society. San Francsico, California, September 10-14, 2006.
5. Helstrom, Carl. Probability and Stochastic Processes for Engineers. Macmillan Publishing, 1991.
6. B. J. Kuijlaars, E. B. Saff, "Distributing many points on a sphere", Mathematical Intelligencer, v19 n1 (1997), pp. 5-11.
7. S. Miller, J. Heinzel, J. Kuseian, D. Hoffman, E. House, V. Tsourapas, A. Nickens. „A Dynamic Model of a Shipboard PEM Fuel Cell Reformer System with an Integrated Gas Turbine". ASME Turbo Expo 2006: Power for Land, Sea, and Air. Barcelona, Spain, May 8-11.
8. Ruskin, http://www.math.niu.edu/~rusin/known-math/97/disco, 2006
9. Song and X. Ma. "New design approach to ultra-clean diesel fules by deep desulfurization and deep dearomatization. Applied Catalys B: Environmental 41 (2003), 207-238.

10. Strait, P. T. *Probability and Statistics with Applications*. Harcourt Brace Jovanovich Inc. 1989.

Acknowledgments

This work was supported by the Office of Naval Research and the Naval Surface Warfare Center, Carderock Division--Philadelphia.

Optimum reactor design for the phenol production

Goodarz Khodabakhshi, Mohammad Hussein Sayyar
Loughborough university, Loughborough, United Kingdom

1. Introduction

All industrial chemical processes are designed to transform cheap raw materials to high value products (usually via chemical reactions). A 'reactor' in which such chemical transformation take place, has to carry out several functions such as bringing reactants into intimate contact (to allow chemical reactions to occur), providing an appropriate environment (temperature and concentration fields, catalysts) for an adequate time and allowing for the removal of products. In order to reach the optimum reactor design following factors should be considered: cost, product quality, safety, environmental impact, conversion, selectivity, stability and operability [1].

Phenol is an important intermediate for the synthesis of petrochemicals, agrochemicals, and plastics. Examples of using phenols as an intermediate are in the production of Bisphenol A, phenolic resins, caprolactam, alkylphenols, and aniline. As there is considerable interest in using the gas phase catalytic oxidation for phenol, two different equipments for the direct oxidation of benzene in the gas phase were set up. The first route was based on using O_2 as an oxidant in a catalytic membrane reactor using palladium as a catalyst [2] and second route was based on using N_2O as an oxidizing agent and Fe/ZSM-5 as a catalyst [3]. A second set of experiments were carried in an stainless steel plug-flow reactor. Figure 1 schematically shows the membrane and the plug flow reactors.

In the present work, the two stated reactors has been modelled and the effects of geometrical parameters such as reactor length and diameter has been considered together with operational parameters such as inlet flow rate and concentration. The results from the two reactors has been compared. Finally the optimum design for the reactor has been suggested.

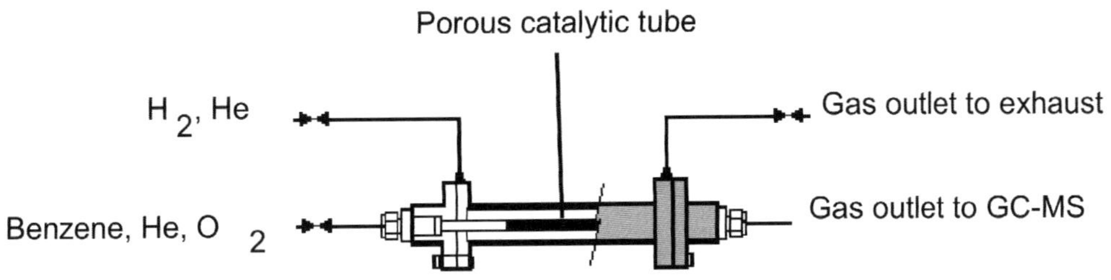

(a)

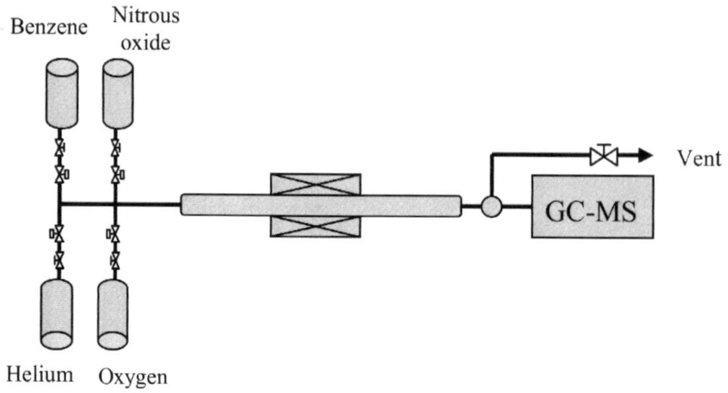

(b)

Figure 1 Reactors
(a) Membrane reactor
(b) Plug flow reactor

2. Mathematical modelling

Using an axial symmetric coordinate system (r, z) the mathematical model describing fluid flow can be written in terms of the conservation of mass and momentum equations. The conservation of mass is written as

$$\frac{\partial \rho}{\partial t} + \nabla \cdot (\underline{V}) = 0 \tag{1}$$

and the conservation of momentum is shown as

$$\rho(\frac{\partial \underline{V}}{\partial t}) + \rho\, \underline{V} \cdot \nabla \underline{V} = \rho \underline{g} + \nabla \underline{\underline{\sigma}} \tag{2}$$

where $\underline{V}$ is the velocity field, ρ is the fluid density and $\underline{g}$ is the body force per unit volume of fluid. The Cauchy stress is given as

$$\underline{\underline{\sigma}} = -p\delta + \underline{\underline{\tau}} \tag{3}$$

where p is hydrostatic pressure, δ is unit second-order tensor (Kronecker delta) and $\underline{\underline{\tau}}$ is the extra stress tensor. Substituting the Cauchy stress from equation (3) into equation (2) leads to the governing equations for the fluid region [4].

For generalized Newtonian fluids the extra stress tensor in the fluid phase is expressed, explicitly, in terms of rate of deformation as

$$\underline{\underline{\tau}} = 2\eta \underline{\underline{D}} \tag{4}$$

where η and $\underline{\underline{D}}$ are the viscosity and the rate of deformation tensors, respectively.

The general equation describing the mass balance is the convection-diffusion equation which is written as:

$$\frac{\partial c_i}{\partial t} + \nabla.(-D_i \nabla c_i + c_i \underline{V}) = R_i \tag{5}$$

where, c_i denotes the concentration of species i, D_i denotes its diffusion coefficient, $\underline{V}$ the velocity vector and R_i denotes the reaction term. The velocity vector obtained by coupling the momentum balance (equation (2)) to the equation system.

In the reaction term, arbitrary kinetic expressions of the reactants and products can be introduced. The expression within the brackets represents the flux vector, where the first term describes the transport by diffusion and the second represents the convective flux. This mass flux vector is:

$$N_i = -D_i \nabla c_i + c_i \underline{V} \tag{6}$$

where N_i is the mass flux vector.

The rate of reaction, R_i, has the following general definition:

$$R_i = -k\, c_i \tag{7}$$

where k is the reaction constant.

As the following reactions occur simultaneously in the rector:

$$C_6H_6 + OH^o \rightarrow C_6H_5OH \qquad k_1 \tag{8}$$

$$C_6H_5OH + OH^o \rightarrow C_6H_4(OH)_2 \qquad k_2 \tag{9}$$

the rates of reaction have been written as:

$$\begin{aligned} R_A &= k_1\, c_A\, c_B \\ R_B &= -k_1\, c_A\, c_B - k_2\, c_A\, c_C \\ R_C &= k_1\, c_A\, c_B - k_2\, c_A\, c_C \end{aligned} \tag{10}$$

where A, B and C represent benzene, the hydroxyl group and phenol, respectively.

2.1. Modelling membrane tube

The membrane reactor was modelled to predict the influence of the geometrical parameters on the production of phenol. Figure 2 schematically shows the reactors geometry and specified boundary conditions. All the simulations were carried out at isothermal conditions. Table 1 shows the numerical values of the parameters used in the simulation.

Table 1 Parameters used in the simulation.

Reaction constant (mol/m^3h)	Benzene inlet velocity (m/s)	Hydrogen inlet velocity (m/s)	Dynamic viscosity (Ns/m^2)	Inlet benzene concentration (mol/m^3)
4.32e-3	0.01	0.001	0.008	10

The conversion of benzene to phenol in a 30mm long reactor with the cross section diameter of 3 mm was modelled. Also, the modelling was carried out when a solid cylinder with 1 mm in diameter is inserted into the tube. The latter case results in an annulus shape geometry.

To validate the constructed model, the results generated by the simulation are compared with the results from the experiments. Table 2 shows the comparison of the simulated and experimental results for the conversion of benzene with and without the inner solid cylinder.

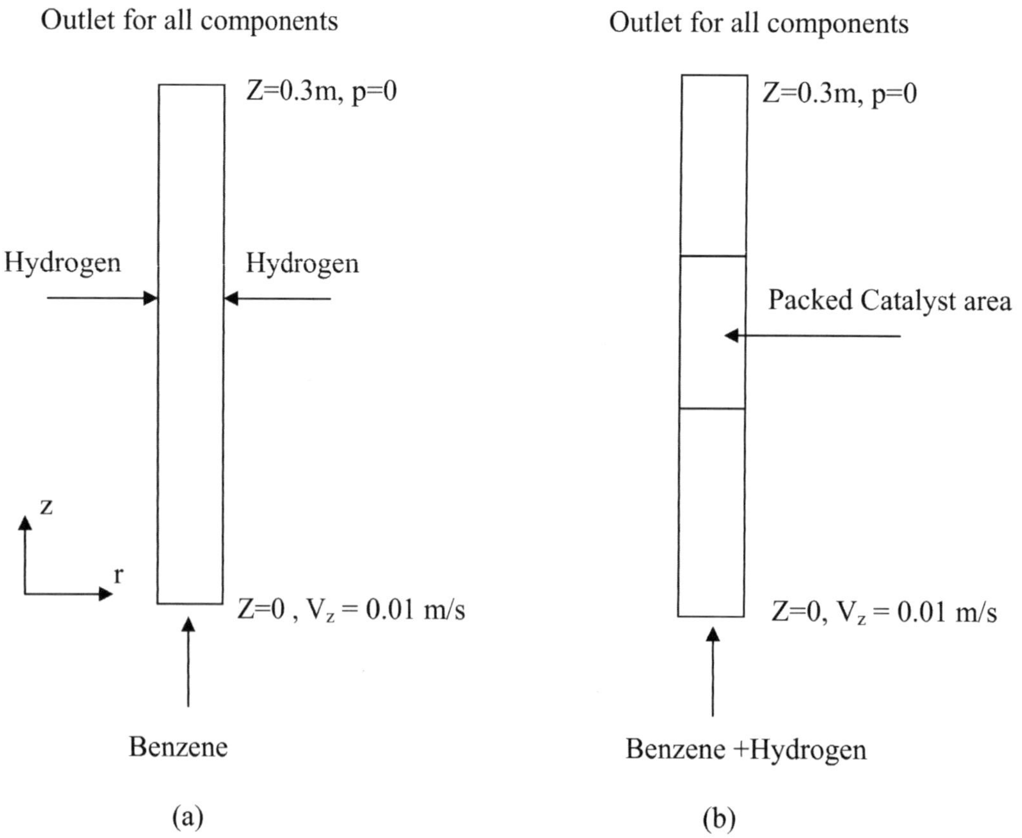

Figure 2 Schematic of the reactor geometry and the specified boundary conditions.
(a) Membrane reactor
(b) Plug flow reactor

Table 2 Experimental and simulation results.

	Phenol (mmol/h)	
	without the inner tube	**with the inner tube of 1 mm diameter**
Experimental	2	3
Simulations	1.8	2.3

Based on the agreement between simulated and experimental results, the proposed model was employed to study the effect of the reactor tube length on the conversion of benzene to phenol. Figure 3 shows the concentration of the produced phenol and conversion of benzene versus the tube length, respectively. As it can be seen in Figure 3, an increase in the tube length results in an increase of phenol production up to the tube length of 60 mm above which the increase in tube length decreases the production of phenol. This length is considered to be the optimum length at which the production of phenol is maximised.

For the annular geometry, the effect of increasing the inserted solid cylinder diameter on the conversion of phenol has been investigated. Figure 4 shows the conversion of benzene versus the annulus thickness (the area between the tube and inserted solid cylinder). Increasing the diameter of the inserted solid cylinder results in narrower annulus which increases the benzene conversion. FEM LAB was used to carry out the simulations.

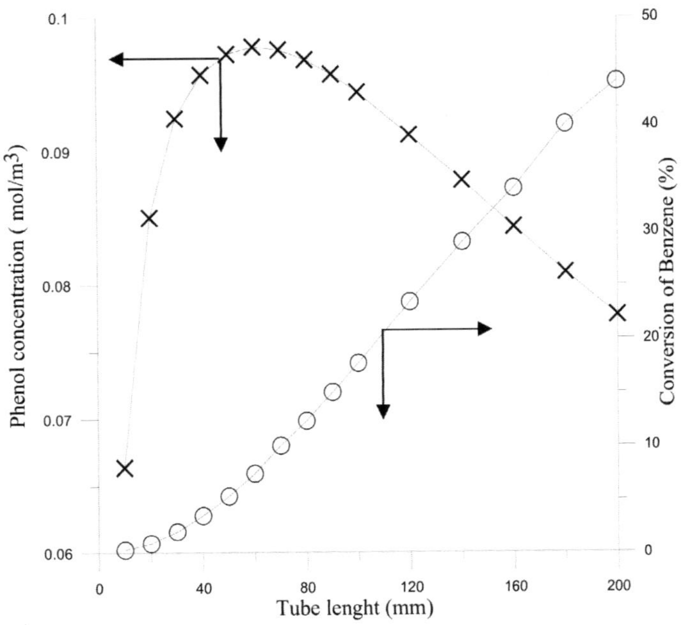

Figure 3 Concentration of phenol (X) and conversion of benzene (O) versus the tube length

(Tube diameter = 3 mm)

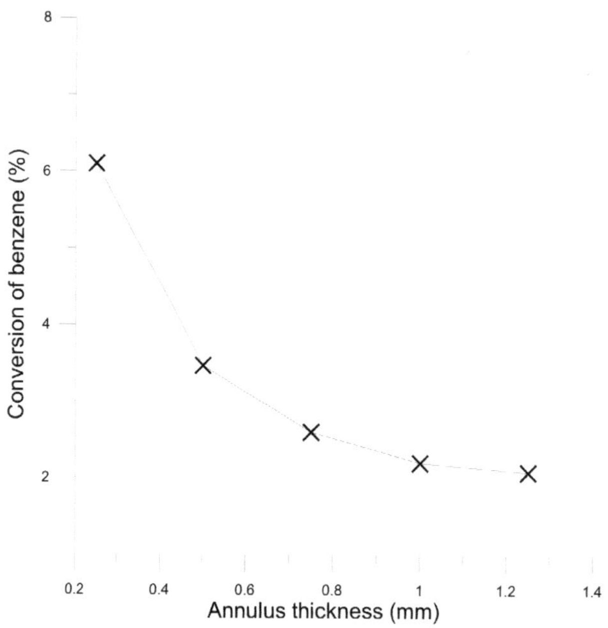

Figure 4 Conversion of benzene versus the annulus thickness
(Tube length = 30 mm)

The effects of changing the flow rate and the composition of the reaction components on the product have been investigated. As it is shown in Table 3, decreasing the flow rate results in higher phenol (product) concentration. Experimental results suggest the value of 0.01 m/s as an optimum inlet velocity. As it can be seen in Table 3, the inlet velocities lower than 0.01 m/s do not change the product concentration dramatically.

Table 3 Experimental and simulation results (Tube length = 30 mm, diameter = 3 mm).

Inlet velocity (m/s)	**0.0001**	**0.001**	**0.01**	**0.1**	**1**
Phenol concentration (mol/m^3)	**0.112**	**0.1106**	**0.092**	**0.037**	**0.0045**

The simulations results show that increasing the inlet concentration of the reactants result in higher concentration of the product (phenol). Although, this factor increases the product concentration but also increases the risk of reaching flammability point and explosion in the system as well. On the other hand, returning (recycling) the benzene which is not converted to phenol from the outlet to the feed to be used again is an expensive process. Therefore, Increase the benzene concentration in the feed will result in an economically inefficient system. As a conclusion, the flow rate and concentration are not proper factors to be optimized just by the mathematical simulation due to the experimental limitations and considerations.

2.2. Modelling plug flow reactor

The second approach to produce phenol is using the plug flow reactor. In this approach the reactants flow in together from the same side and the reaction starts when they meet the catalyst which is packed in the middle of the tube. Mathematical model has been constructed based on the data provided by the experiments. Table 4 shows the experimental data.

Table 4 Parameters used in the simulation [5].

Reaction constant, k_1 (mol/m^3h)	inlet velocity (m/s)	Inlet hydrogen concentration (mol/m^3)	Reaction constant, k_2 (mol/m^3h)	Inlet benzene concentration (mol/m^3)
9.3e-3	0.028	4.54	1.85e-4	45.4

The experiments show the value of 1.86 (mol/m^3) for the concentration of phenol as the simulations value is 2.12 (mol/m^3). The agreement between simulated and experimental results confirmed the reliability of the constructed model. Figure 5 shows the concentration of phenol across the length of the reactor tube.

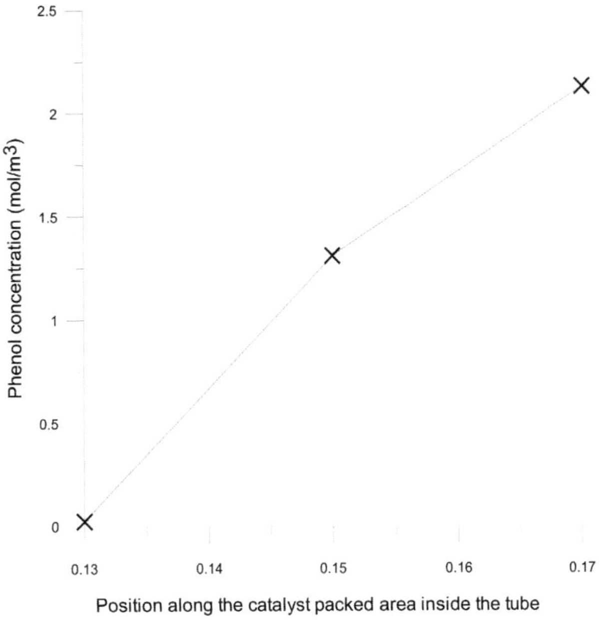

Figure 5 Concentration of phenol along the tube length (Catalyst packed area length = 4 mm, diameter = 3 mm)

3. Conclusion

The modelling shows longer tubes increase the conversion of benzene but there is an optimum length because of preventing side reactions. The tube length of 60 mm results in the maximum product. Also the modelling results show higher productivity using a solid tube inside the membrane. Due to the experimental restrictions the operational parameters can not be reliable solely. As the modelling results shows, the plug flow reactor results in higher amount of product at the out let.

References

1. Ranade.V, (2002), Computational flow modelling for chemical reactor engineering, academic press.

2. Niwa, S.; Eswaramoorthy, M.; Nair, J.; Raj, A.; Itoh, N.; Shoji, H.; Namba, Y.; Mizukami, F., **(2002)** *J. Sci.* 295.

3. Y. Liu, K. Murata and M. Inaba, *Catal. Comm.*, 6 **(2005),** pp. 679-683.

4. Nassehi,V., **(2002)**, Practical aspects of finite element modelling in polymer processing. John Wiley , Chichester.

5. Reitzmann.A, Klemm.E, Eming.G., *Chem..Eng. J.*, 90,**(2002)**, pp.149-164.

Singularities in Reactive Separation Process Problems

Lakshmi Sridhar, Chemical Engineering, University of Puerto Rico, Mayaguez campus, P.O. Box 9046, Mayaguez, PR 00681 and Gerardo Ruiz, chem engg, UPR MAyaguez, PO Box 9046, uprm, Mayaguez, PR 00681.

In this article we present some results for isothermal isobaric reactive separation process problems. We also demonstrate that even in isothermal isobaric reactive separation processes which is probably the least nonlinear of all reactive separation processes we get nonlinear phenomenon such as Hopf bifurcations. While it has been shown that Hopf bifurcations are impossible in isothermal CSTR problems involving the MTBE and TAME reactions, and also in non-reactive flash problems, we demonstrate in this paper that isothermal reactive flash processes involving both MTBE and TAME mixtures exhibit Hopf bifurcations. This shows that instabilities and oscillations can occur even in isothermal reactive separation systems and are not necessarily due to multiple stages. Additionally, we show that the Rachford-Rice procedure can be extended to reactive systems.

During the last decade there has been a tremendous interest in the field of reactive distillation. A review of the various models used in reactive distillation can be found in Taylor and Krishna(2000). Of special interest is the existence of multiple steady-states in these problems, since the combination of separation and reaction can in principle introduce the nonlinearity that can cause multiplicity. Multiple steady-states in reactive distillations was demonstrated by several workers (Jacobs and Krishna; 1993, Nijhuis, Kerkhof and Mak, 1993, Bravo, Pylahathi and Jarvelin, 1993, Hauan, Hertzberg and Lien; 1995, 1997; Hauan, Schrans and Lien, 1997; Sneesby, Tade and Smith; 1998a, 1998b; Eldarsi and Douglas, 1998; Mohl Kienle and Gilles, 1998, 1999; Rapmund, Sundmacher and Hoffmann, 1998; Higler, Taylor and Krishna, 1999; Guttinger and Morari (1999a, 1999b); Chen, Huss, Doherty and Malone, 2002). The most commonly investigated situations include the MTBE synthesis in the Jacobs-Krishna (1993) column configuration and the TAME Synthesis in the column of Mohl et. al (1999). The multiple steady-states for these two columns were investigated by Chen, Huss, Doherty and Malone, (2002) who conclude that multiplicities are lost for high values of Da for TAME, while the opposite is found for MTBE. This conclusion, however is specific to the column configuration descrbed in Jacobs-Krishna (1993) and Mohl et al (1999). Rodriguez et al (2002, 2004) discuss causes for the existence of multiple steady-states in binary and ternary systems. The most important reactive separation process problems where multiplicity exists such as MTBE and TAME processes involve more than three components. In order to understand what causes multiplicity in these problems one must look at the simplest reactive separation process problem involving the MTBE and TAME mixture and hence we are motivated to look at the isothermal reactive flash problem. Mohl et al (1999) prove that isothermal CSTR problems involving the MTBE and TAME reactions do not exhibit Hopf bifurcations while, on the other hand, for non-reactive isothermal flash processes involving homogeneous mixtures Hopf bifurcations are impossible (Lucia 1986). However we demonstrate that isothermal reactive flash processes involving both TAME and MTBE exhibit Hopf bifurcations and that is one of the important contributions of this paper. This paper is organized in the following manner. First, a brief description of the isothermal reactive flash process is given along with the equations involved.. We then demonstrate the existence of Hopf bifurcations in the isothermal reactive flash processes involving both the MTBE and TAME mixtures. Dynamic simulations are performed demonstrating the existence of limit cycles that are a characteristic feature of problems with Hopf bifurcations and the behavior of these Hopf bifurcation points with temperature and pressure variations are presented.

Optimal Design of Heat-Integrated Separation and Refrigeration Systems

Robin Smith, Megan Jobson, and **Sonia Farrokhpanah**. Centre for Process Integration, The University of Manchester, Sackville Street, PO Box 88, M60 1QD, Manchester, United Kingdom

Refrigeration systems in gas separation plants account for a large contribution to the capital and operating costs. This paper addresses the interactions between the separation and refrigeration systems. Designs must provide the optimal sequence of separations, types of separation operation and associated operating conditions, to separate a feed into the desired products, together with the design of the refrigeration system. This paper focuses on simultaneous design of the refrigeration and separation systems. Optimisation variables, regarding refrigeration system, include the number of evaporation pressure levels, multiple rejection levels, heat loads on each refrigeration level, partition temperature in cascaded cycles, composition in mixed refrigerant systems, etc. Heat pumps can also be included. Stochastic methods have been adopted for optimizing the formulated optimisation problem. Case studies will focus on different separation sequences coupled with various refrigeration techniques. Both pure and mixed refrigerant cascaded refrigeration cycles are considered.

A Dynamic Model for a Pump Hydraulic System

Richard Carranza, -, Houston, TX 77042

A dynamic model is presented for a pump hydraulic system. The model is based on a modified form of the Bernoulli Equation. The Bernoulli Equation is adjusted, using discret differences, such that it is redefined as a function of time. The Equation is solved, stepwise, for a given time period using a compiled fortran program. The output from the program is then imported to a spreadsheet and represented in graphical form.

Web Page: www.richardcarranza.com

Reliable solution algorithms for global optimization, such as Interval Analysis, had been introduced to find all the stationary points (solutions) within given domain. However, performance remains as a challenging problem. This presentation introduces a new solving strategy by connecting Generalized-Reduced-Gradient algorithm and Interval Analysis. The new method, which is based on bi-level iterative scheme, has the potential to reduce the number of variables as well as simplify the problem specifications without losing the promise of finding all solutions within given domain.

The problem for bi-level iteration is that it is known for its jacobian complexity as well as restrained solution path. Thus, one may select single-level iteration depending on the problem specifications. It has been found, however, that the performance for global solution methods depend as much as (or even more) on the problem dimensionality than on the problem complexity. Meanwhile, simple or even linear constraints exist for majority optimization problems. Thus, if inner-level scheme can be defined not only simple but also reducing the problem dimension significantly, then applying global optimization for certain bi-level iteration problems can provide both reliability and efficiency.

Details on how to modify interval analysis with bi-level iteration is discussed and preliminary tests have been done by this presentation. This is also the first time a rigorous bi-level iterative modification on interval analysis is addressed.

Investigation of Economic Penalty Due to Measurement Bias in Real Time Optimization

Peyton C. Richmond, Chemical Engineering Department, Lamar University, 4400 MLK Blvd, Beaumont, TX 77710

Real Time Optimization (RTO) is attractive to chemical manufacturers because it can both improve their profitability and reduce unwanted emissions simultaneously. Even small increases in facility profitability can be worth hundreds of thousands of dollars per year. However, potential RTO benefits may not be realized due to economic penalties caused by process measurement bias errors. These penalties have not been thoroughly explored in previous research about the industrial effectiveness of RTO applications. This research will investigate the economic penalty due to biased plant measurements and develop a procedure to determine when bias error is significant to a particular RTO design.

Synergistic Approach to Real Time Optimization of Modern Industrial Processes

Ravi Nath, Sanjay Sharma and Anil Gokhale
Honeywell Process Solutions

paper 27c

AIChE Spring National Meeting
Houston, TX
April 23, 2007

Agenda

- Introduction
- MPC approaches
- RTO approaches
- Synergistic approach
- An example
- Conclusions

Introduction: Reg. Control

- Regulatory Control
  - to maintain a controlled variable at a desired value by adjusting a manipulated variable.
- CV
- MV
- Algorithm
  - 3 parameters

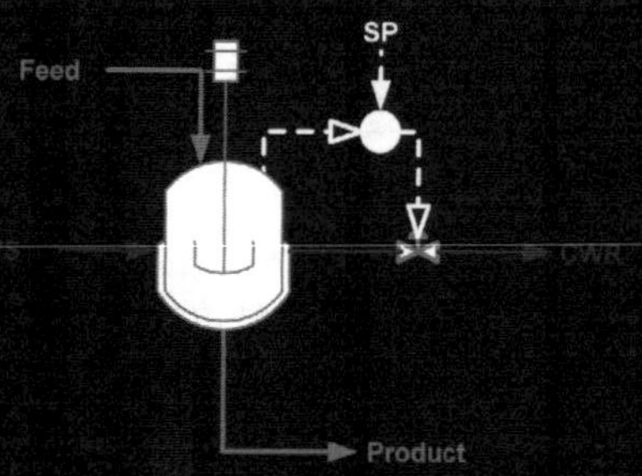

Introduction: Reg. Controls (2)

On the whole, regulatory controls work well

- Fast execution cycle (sub second) and feedback.
- PID problems
  - Interactions between loops
  - Process behavior is highly non-linear
  - Inverse response
  - Dead time

Introduction: MPC

- Model Predictive Control (MPC)
  - to maintain controlled variables at desired values (or within range) by adjusting manipulative variables.
- CVs
- MVs
- DVs
- Sub-models
  - Linear dynamic
    - Empirically identified

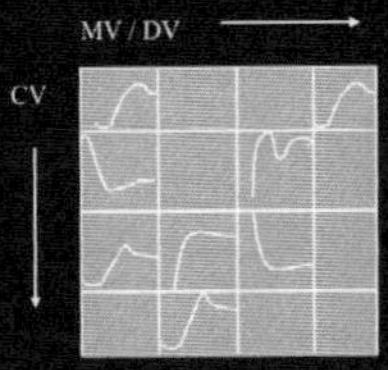

Introduction: MPC (2)

- Feedback is essential

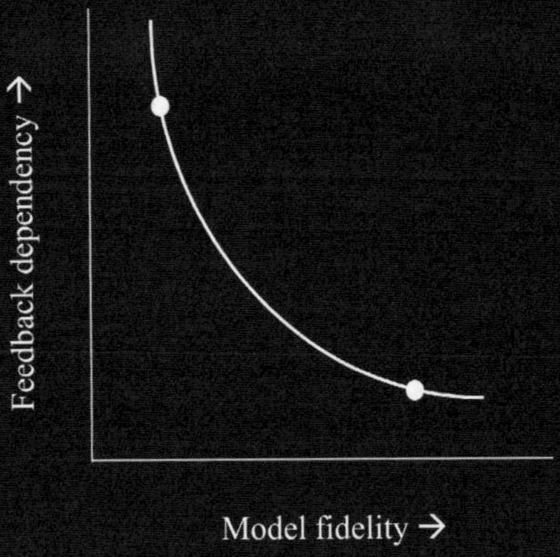

Introduction: Optimization

- Real Time Optimization (RTO)
  - to determine the optimum and to move the process towards the optimum by adjusting independent variables.
- Objective function
- DOF
- Model
  - Process model is essential
  - Model scope
  - Model fidelity
  - Execution frequency

Introduction: Optimization (2)

- Model scope versus "true" benefits

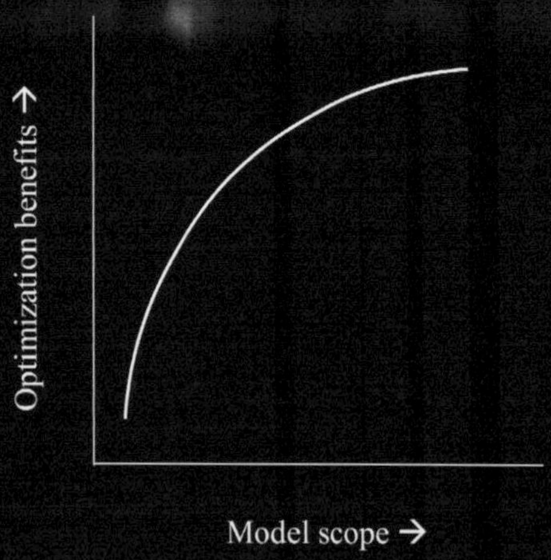

Introduction: Optimization (3)

- Model complexity versus resource requirements

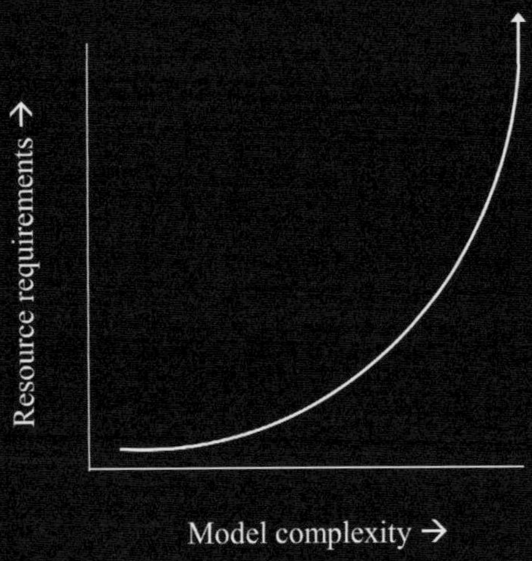

AIChE Spring National Meeting, April 23, 2007 9

Introduction: Control and Optimization

- Control and optimization must cooperate
  - Optimization must yield when control problem appears
  - Optimization must drive otherwise

AIChE Spring National Meeting, April 23, 2007 10

Introduction: Summary

- Feedback is essential both for control and optimization
  - because models are not perfect
- Faster execution
  - makes up for divergence between process and model
- Larger optimization scope
  - for more "true" benefits
- Simpler models
  - for maintainability and ease of development
- Cooperation between Control and optimization
  - for continuous benefits

MPC approaches

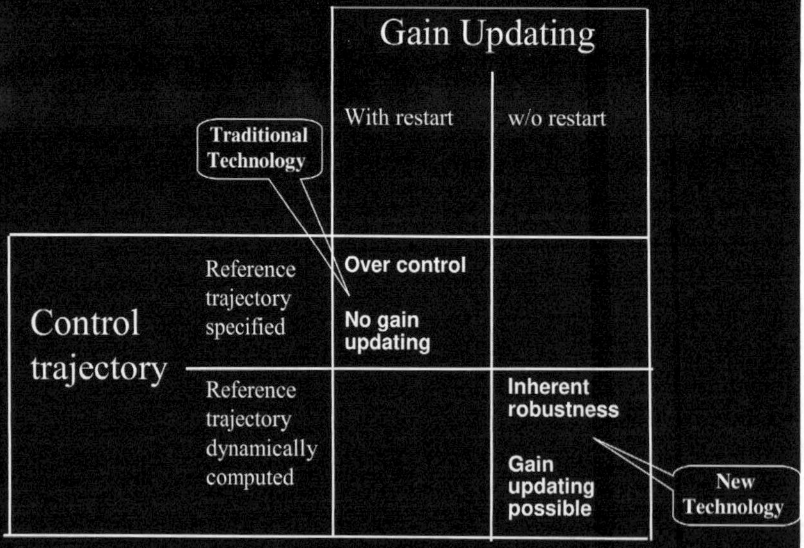

RTO Approaches: Optimization model

- Optimization needs a model.
- Model is an approximation of the process.
- Important aspects
  - Data reconciliation: **Least squares versus biasing**
  - Model type: **Linear versus Non-linear versus successive linear**
  - Model dynamics: **Dynamic versus steady state**

RTO approaches

		Temporality: Steady state model	Temporality: Dynamic model	
Linearity	Linear model (LP)	Algebraic eqn. Fast execution Wait for SS	Differential eqn. Fast execution No wait	Synergistic Technology
	Seq. Linear model (SLP)	Algebraic eqn. Fast execution Wait for SS	Differential eqn. Fast execution No wait	Synergistic Technology with gain updating
	Non-linear model	Algebraic eqn. Slow execution Wait for SS (Traditional Technology)	Differential eqn. Slow execution No wait	Emerging Technology

Synergistic Solution: MPC (trade name Profit Controller)

Unique features are

- Funnel design / Range Control algorithm
  - adds inherent robustness
  - minimum move solution
  - Unified control and optimization
- Parametric models
  - On the fly gain updating
- Controller level optimization
  - Linear and quadratic terms, in terms of both MVs and CVs

AIChE Spring National Meeting, April 23, 2007 15

Synergistic Solution: MPC: Range Control

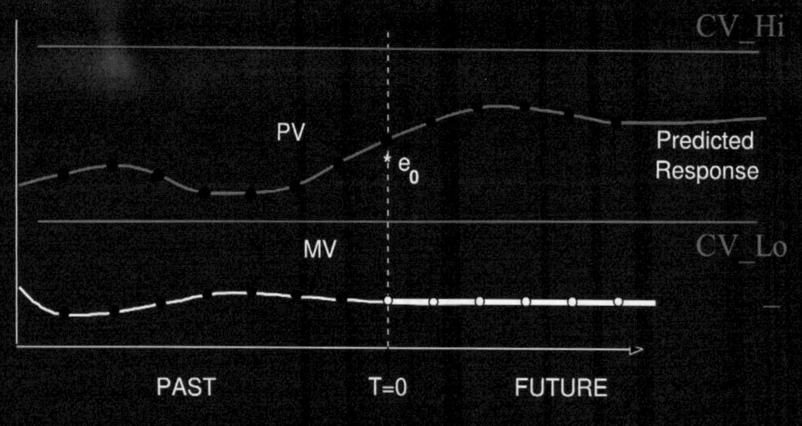

AIChE Spring National Meeting, April 23, 2007 16

Synergistic Solution: MPC: Funnel design

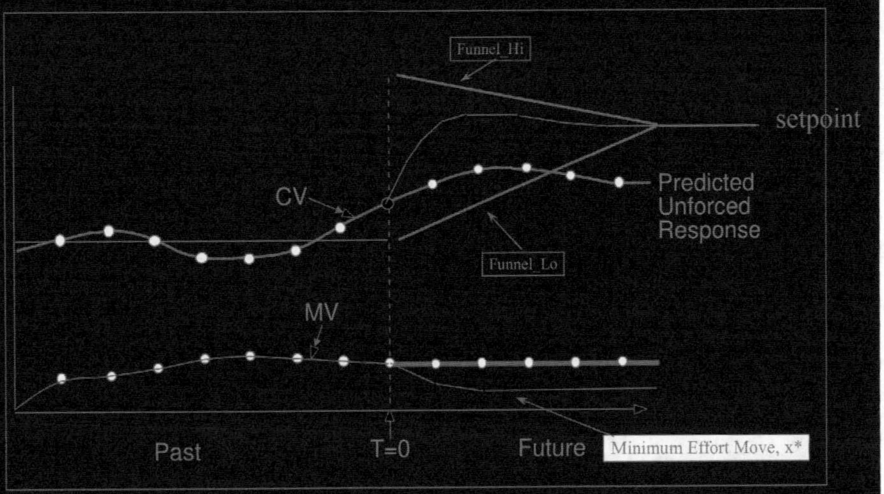

AIChE Spring National Meeting, April 23, 2007 17

Synergistic Solution: MPC: Controller level optimization

- **Why Optimize?**
  - **DOF, when Number of MVs > Number of constrained CVs**
- **Objective function**
  - **Linear terms (PVO)**
  - **Quadratic terms (QP)**

$$\text{MAX } J = 30 * CV1 - 20 * MV1$$

Coefficients

$$\text{MIN } J = 2 * (MV2 - 5)^2 + (CV3 - 7)^2$$

Desired values

AIChE Spring National Meeting, April 23, 2007 18

Synergistic Solution: MPC: Unified Control and Optimization

Control problem is to maintain the CVs within limits.
Optimization problem is to maintain the MVs at their optimum values.

Unifies control and optimization

- maintains CVs within the Control Time Horizon
- maintains MVs within the Optimization Time Horizon
- strives to minimize MV movement.

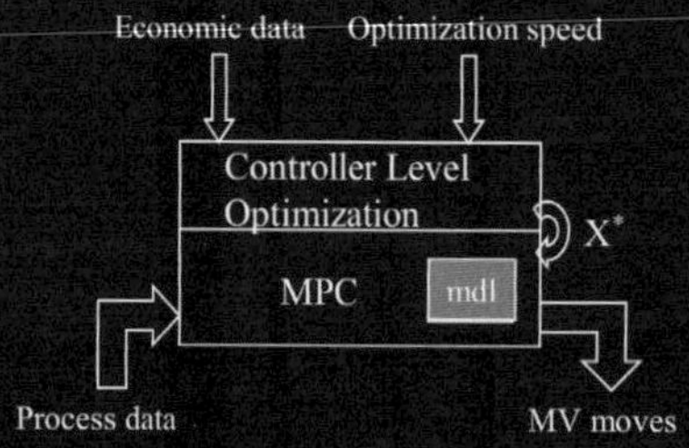

Synergistic Solution: Multiple MPC

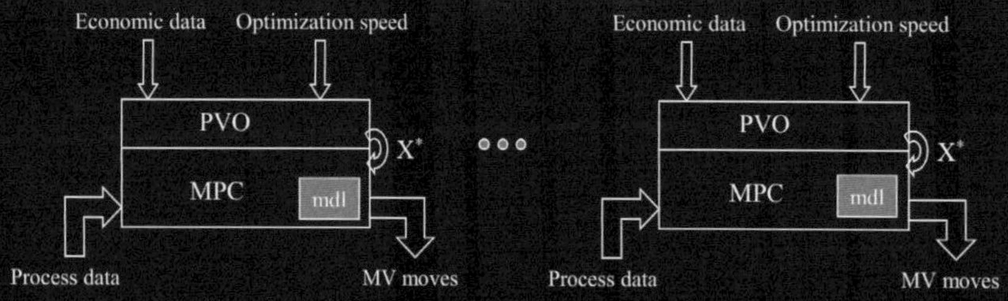

Too Many Objective functions … may be conflicting

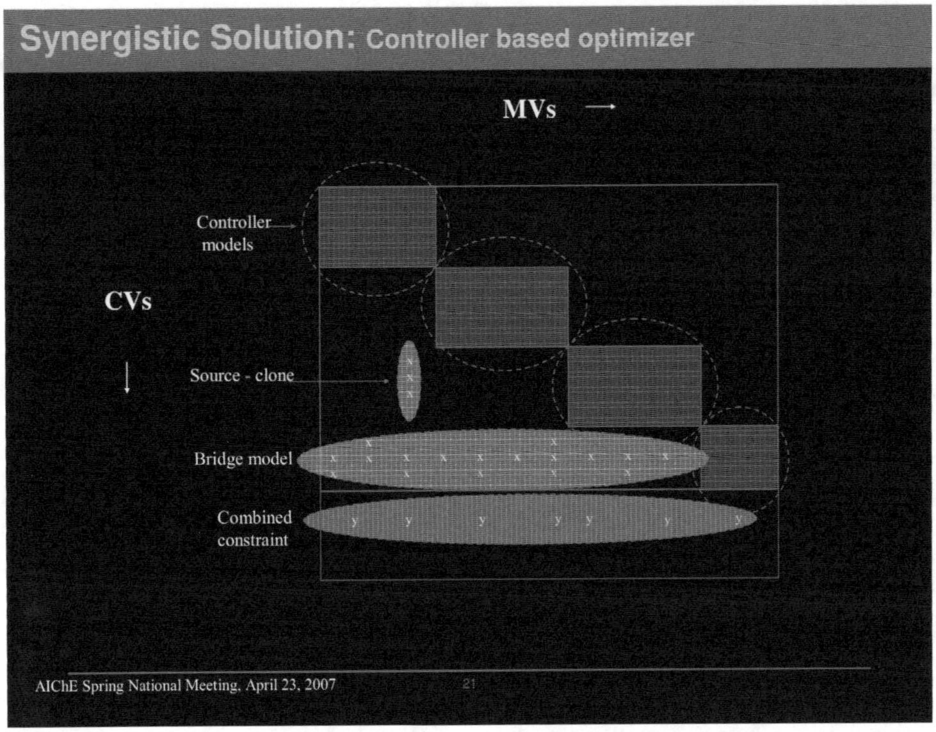
Synergistic Solution: Controller based optimizer
MVs
CVs
Controller models
Source - clone
Bridge model
Combined constraint
AIChE Spring National Meeting, April 23, 2007
21

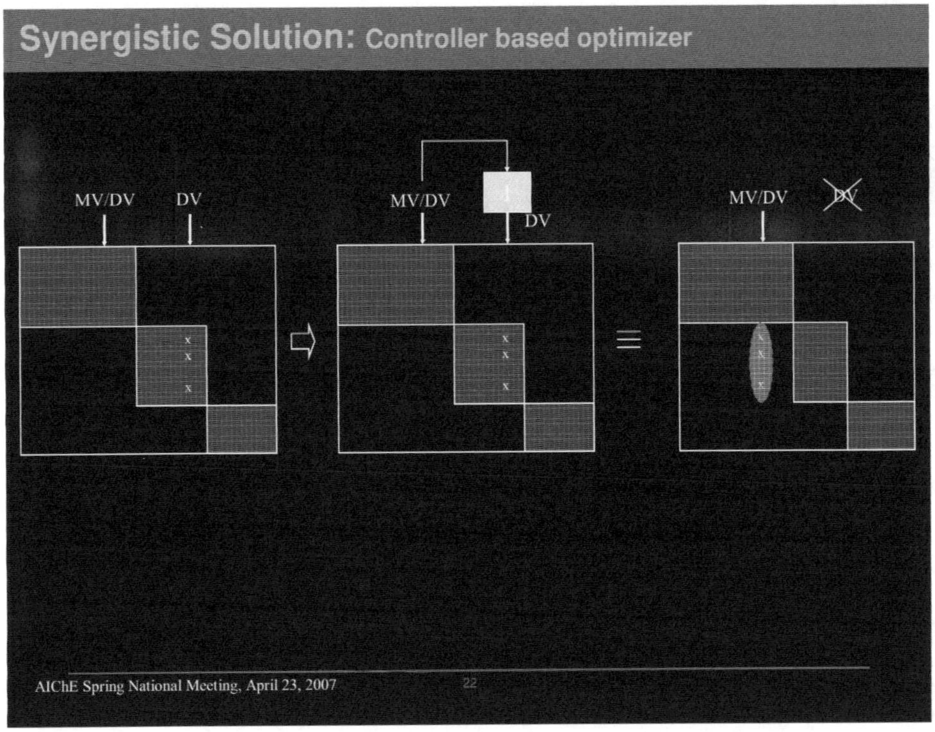
Synergistic Solution: Controller based optimizer
MV/DV
DV
MV/DV
DV
MV/DV
DV
AIChE Spring National Meeting, April 23, 2007
22

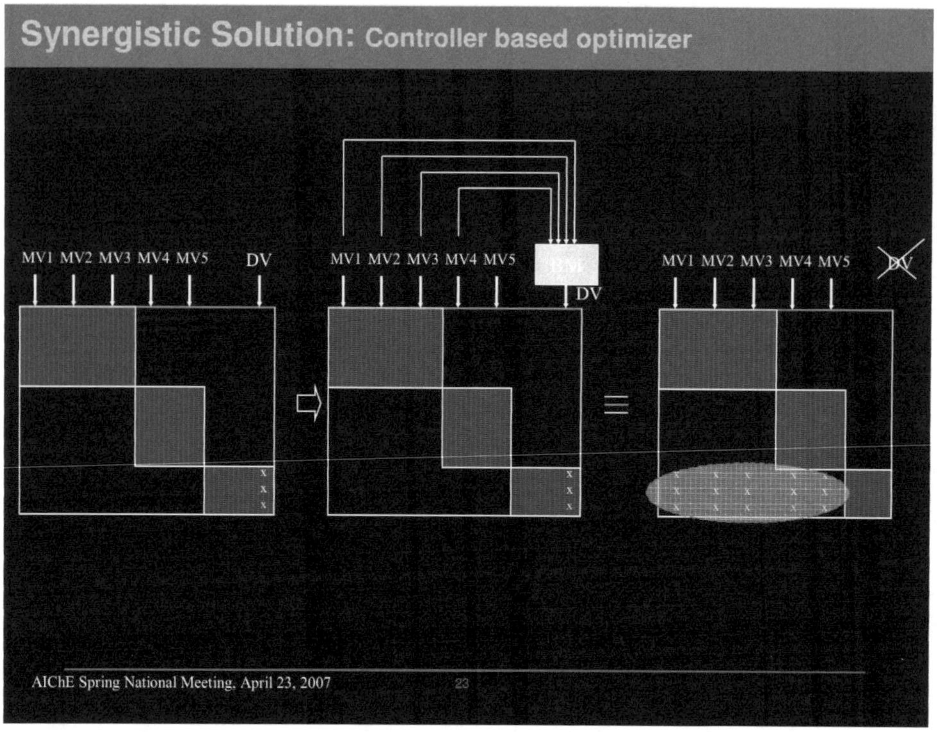
Synergistic Solution: Controller based optimizer
MV1 MV2 MV3 MV4 MV5
DV
MV1 MV2 MV3 MV4 MV5
DV
MV1 MV2 MV3 MV4 MV5
DV
x
x
x
AIChE Spring National Meeting, April 23, 2007
23

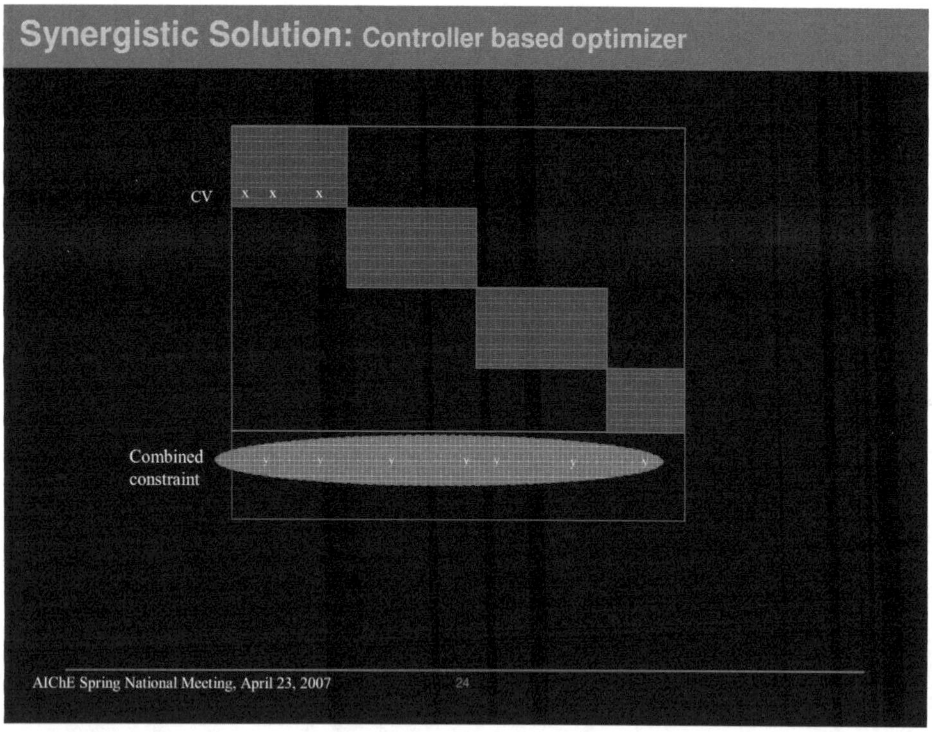
Synergistic Solution: Controller based optimizer
CV
x x x
Combined
constraint
AIChE Spring National Meeting, April 23, 2007
24

Synergistic Solution: Controller based optimizer

Non-linearities are addressed by gain updating at a relatively fast frequency.

- External non-linear models are exercised to compute controller and optimizer gains at the new operating point
- New controller and optimizer gains are updated without re-initialization

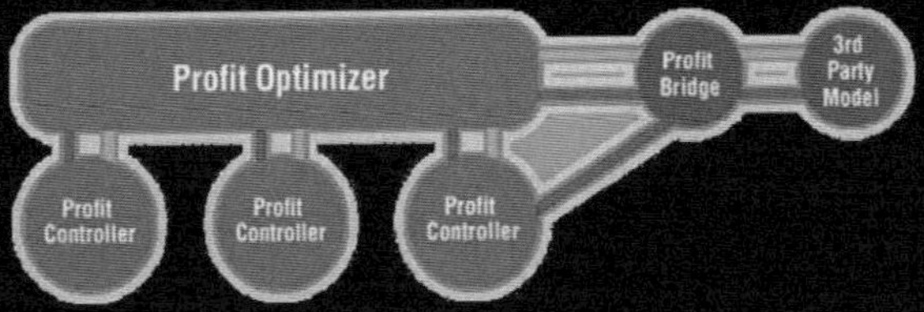

Synergistic Solution: RTO (trade name Profit Optimizer)

- MODELING
  - **Single model for control and optimization**

  versus
  - **Separate models for control and optimization**
- STATE ESTIMATION
  - **Dynamic model, scheduled execution at controller frequency**

  versus
  - **Steady State model, wait for steady state**
- OPTIMUM STATE
  - **Successive LP**

  versus
  - **NLP**
- DYNAMIC PATH
  - **Minimum move feasible trajectory**

  versus
  - **Large Moves / CV violation tradeoff**

Significantly reduced model development and maintenance cost.

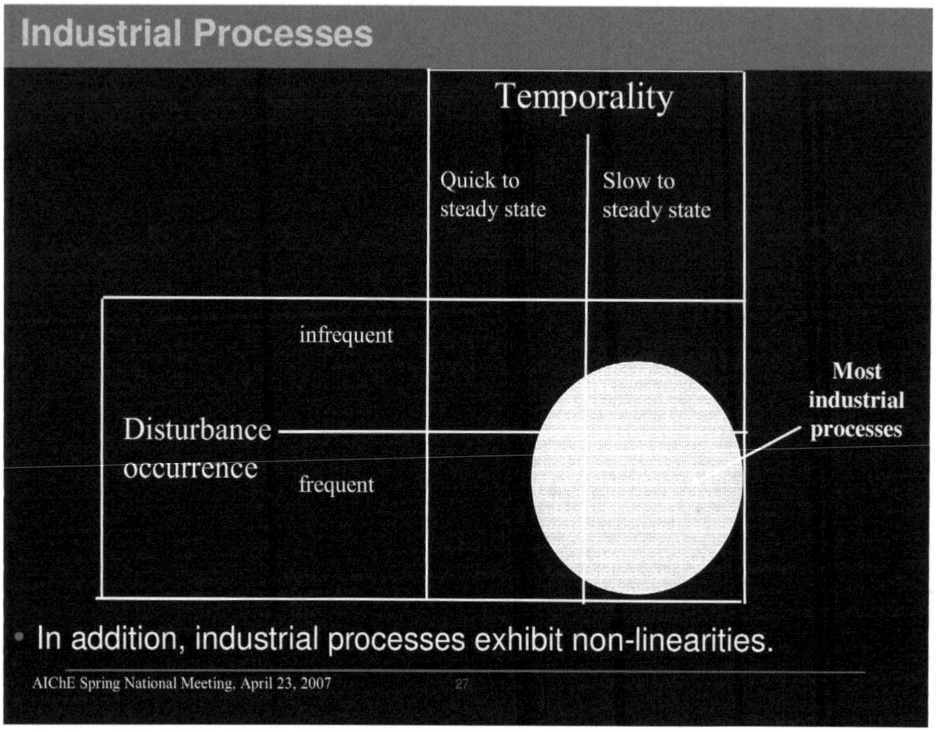
Industrial Processes
Temporality
Quick to steady state
Slow to steady state
infrequent
Disturbance occurrence
frequent
Most industrial processes
In addition, industrial processes exhibit non-linearities.
AIChE Spring National Meeting, April 23, 2007
27

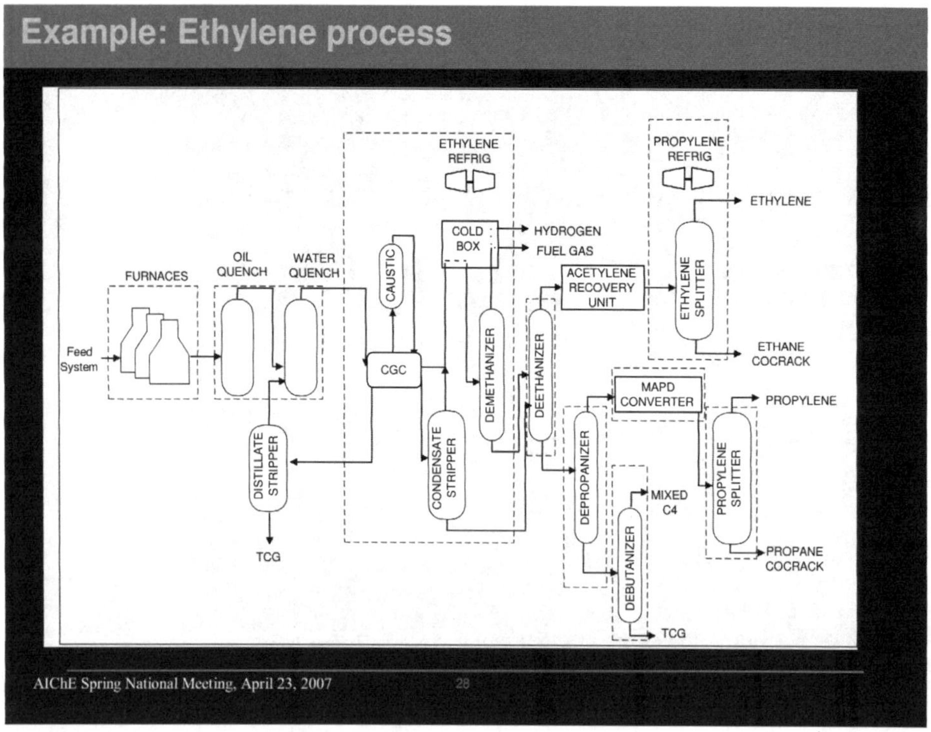
Example: Ethylene process
ETHYLENE REFRIG
PROPYLENE REFRIG
ETHYLENE
COLD BOX
HYDROGEN
FUEL GAS
OIL QUENCH
WATER QUENCH
FURNACES
CAUSTIC
ACETYLENE RECOVERY UNIT
ETHYLENE SPLITTER
Feed System
CGC
DEMETHANIZER
DEETHANIZER
ETHANE COCRACK
MAPD CONVERTER
PROPYLENE
DISTILLATE STRIPPER
CONDENSATE STRIPPER
DEPROPANIZER
MIXED C4
PROPYLENE SPLITTER
DEBUTANIZER
PROPANE COCRACK
TCG
TCG
AIChE Spring National Meeting, April 23, 2007
28

Example: Implementation summary

- Scope:
  - Cracking furnaces
  - Cracked Gas compressor
  - DC1, DC2, C2S, DC3, C3S, DC4 and Converters
- Schedule:
  - 9 month duration
- Execution frequency:
  - MPC: ½ to 1 minute
  - RTO: 1 minute
  - Gain updating: 5 minutes

AIChE Spring National Meeting, April 23, 2007 29

Example: Production increase

Normalized Feed and Production rate trends

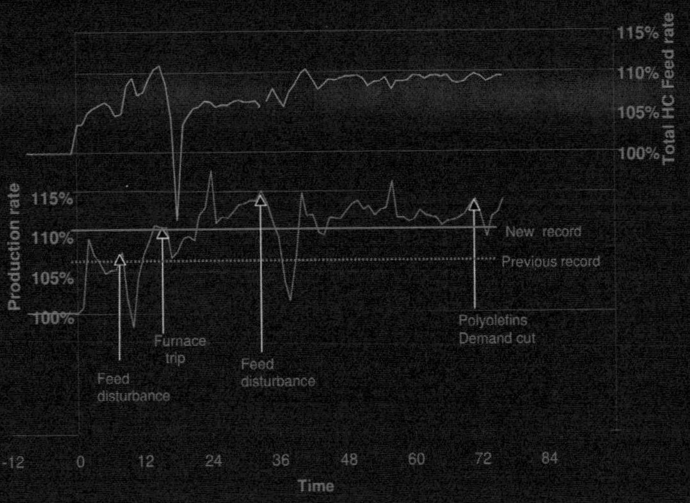

AIChE Spring National Meeting, April 23, 2007 30

Conclusions

- A synergistic approach to optimization that fully utilizes control models is possible.
- The synergistic approach accommodates process non-linearities in an efficient manner.
- The synergistic approach significantly reduces the development, implementation and maintenance effort .
- The synergistic approach has worked well for industrial processes especially those with slow dynamics and frequent disturbances.

References

Bartusiak, D. and K. Sheth, "**Maintaining Realtime Optimization Applications**", 14th Annual Ethylene Producers' Conference, March 2002.

Boell, M, R. Vasbinder and , N. Castelijns, "**Optimising Ethylene Plant Production Using Model Predictive Control and Dynamic Real Time Optimisation**", 16th Annual Ethylene Producers' Conference, March 2004.

Lu, J. and J. Escarcega, "**Introduction to Dynamic Economic Optimization**", Honeywell Internal Technology Document (1998).

Nath, R., Alzein, Z., Pouwer, P., and M. Laiseca, "**On-line Dynamic Optimization of an Ethylene Plant using Profit Optimizer**", NPRA Computer Conference, November 1999

Nath, R., Alzein, Z., Pouwer, P., and R. Voetsch, "**Dynamic Real-time Optimization and Process Control of Twin Olefins Plants at DEA Wesseling Refinery**", 13th Annual Ethylene Producers' Conference, April 2001.

Ng, C., Ang, K., Khair, M, Yap, P., Halberg, M. and R. Nath, "**On-line Dynamic Optimization of PCS Olefins I**", 14th Annual Ethylene Producers' Conference, March 2002.

Verne, T., B. Jerrit, and P. Bester, "**Unique approach to real time optimization**", Hydrocarbon Processing, 78(3), p 53, (1999).

OPTIMIZATION TOOLBOX FOR MODELING OF MULTIPLE REACTION SYSTEMS

V. SREENIVAS
Scientist
Cell for Industrial Safety and Risk Analysis
Chemical Engineering Division
C.L.R.I., Chennai – 600 020, INDIA.

OVERVIEW

- Introduction: Single / Multiple Reaction Systems
- Modeling of Runaway Reactions – Limitations of traditional approach
- Revised Methodology for Modeling of Multiple Reaction Systems
- Case Study: Di-Tert Butyl Peroxide System
- Applications

Introduction

Multiple Reaction Systems

- Reactions classified as elementary or non-elementary depending on level of correspondence between stoichiometry and rate equation.
- Complex reaction system observed as a single reaction, which in reality is the overall effect of a sequence of elementary reactions.
- Multiple reactions classified as (a) Series, (b) Parallel – competitive and side-by-side.
- Examples: DTBP and MEKPO Systems

Classical Approach to Modeling of Runaway Reactions

$$k = Ae^{-\frac{E}{RT}}$$

$$k^{*} = C_o^{n-1}k = \frac{m_T}{\left(\frac{T_f - T}{\Delta T_{AB}}\right)^n \Delta T_{AB}}$$

$$ln\, k^* = ln\, AC_0^{\,n-1} - \frac{E}{R}\left(\frac{1}{T}\right)$$

$$E = \frac{1.987\times10^{-3}\ln\left(\frac{k_2^*}{k_1^*}\right)}{\left(\frac{1}{T_1}-\frac{1}{T_2}\right)},\ \frac{\text{Kcal}}{\text{mol}}$$

$$AC_o^{n-1} = \frac{k^*}{60}\,\frac{E}{1.987\times10^{-3}}\left(\frac{1}{T}\right)(\text{sec}^{-1})$$

Predicted Self-Heat Rate (Townsend & Tou, 1980):

$$m_T = m_0 \left(\frac{T_f - T}{\Delta T_{AB}} \right)^n e^{\frac{E}{R}\left(\frac{1}{T_0} - \frac{1}{T} \right)}$$

Limitations of applying classical modeling approach to multiple reaction systems

Adiabatic Calorimeter Data

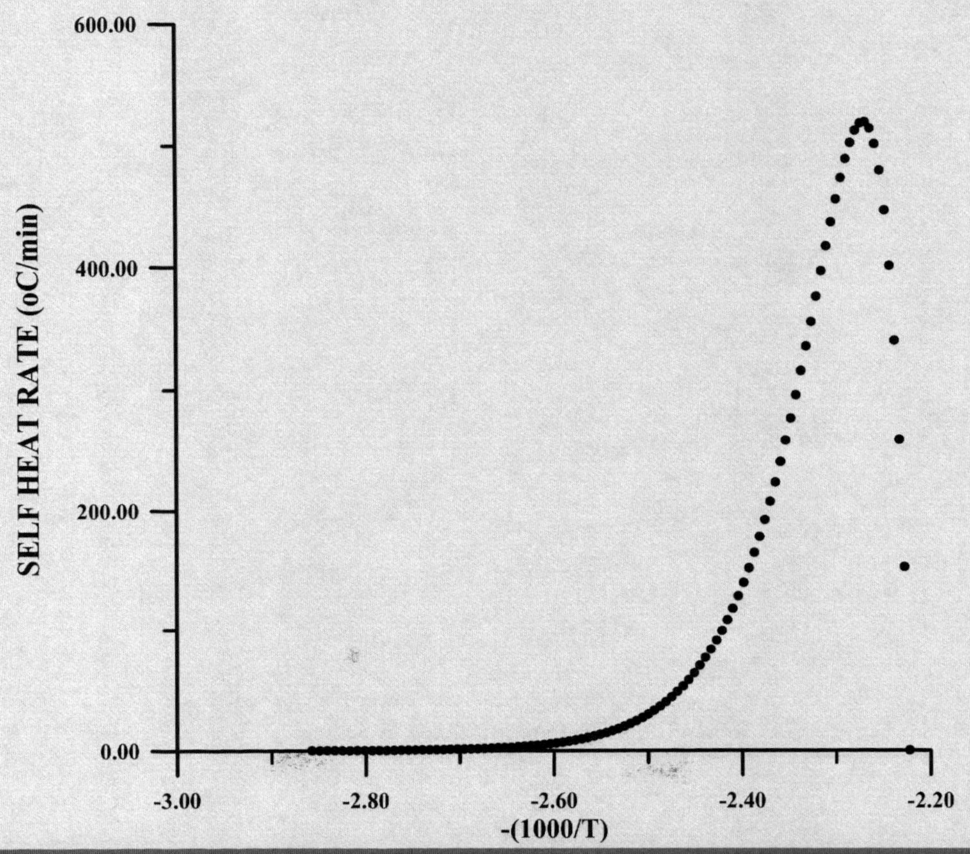

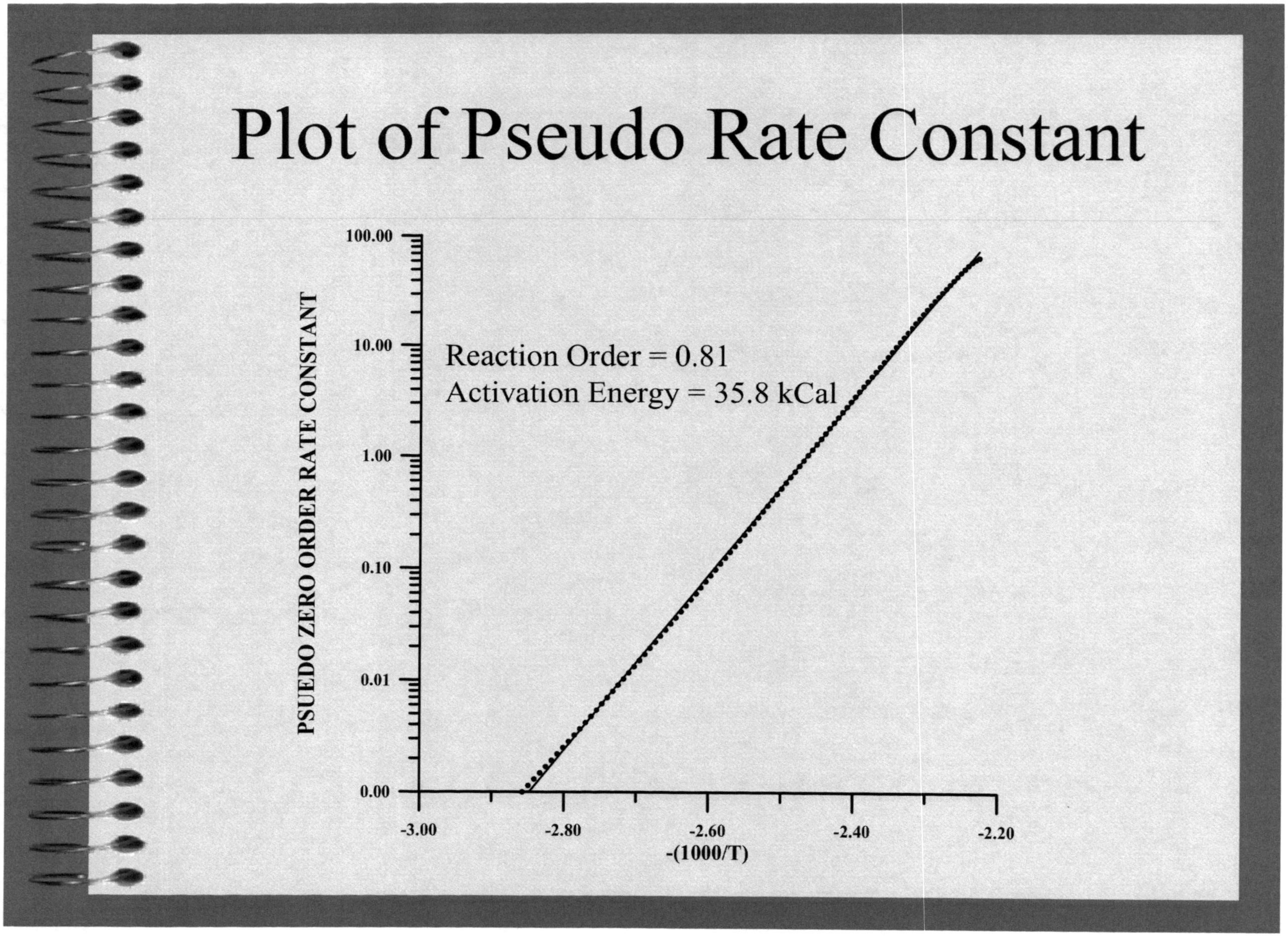
Plot of Pseudo Rate Constant
Reaction Order = 0.81
Activation Energy = 35.8 kCal
PSUEDO ZERO ORDER RATE CONSTANT
100.00
10.00
1.00
0.10
0.01
0.00
-3.00
-2.80
-2.60
-2.40
-2.20
-(1000/T)

Predicted Self-Heat Rate

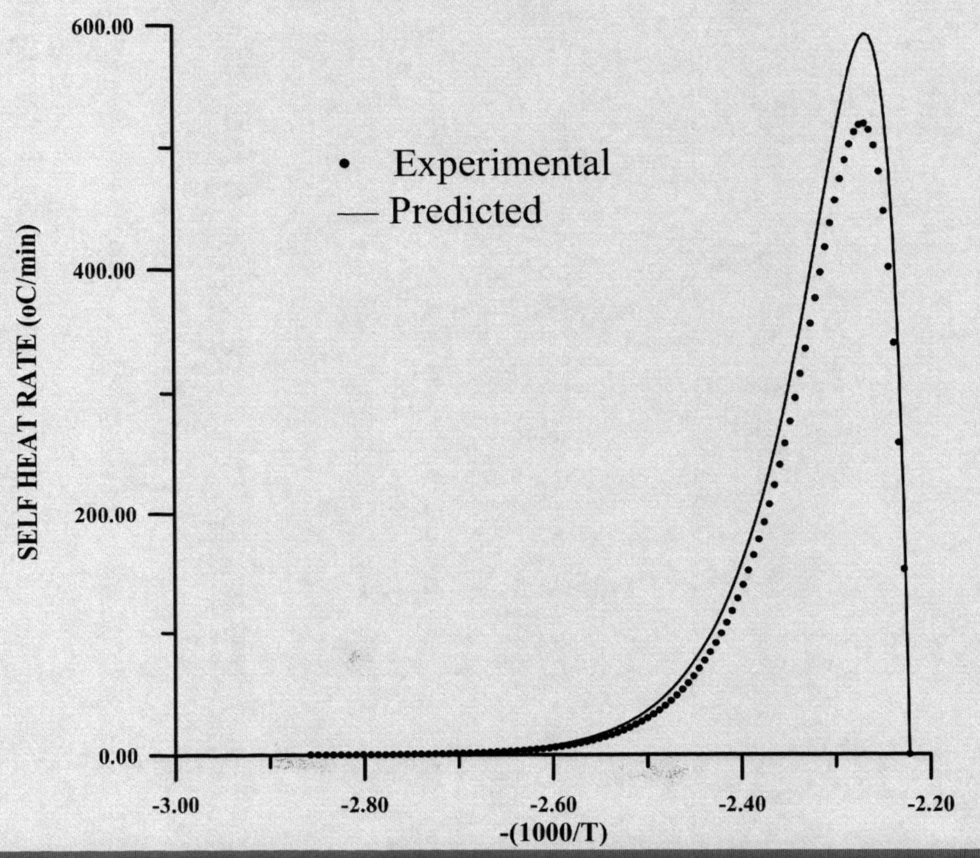

Pseudo Rate Constant Plot (quadruple reaction system)

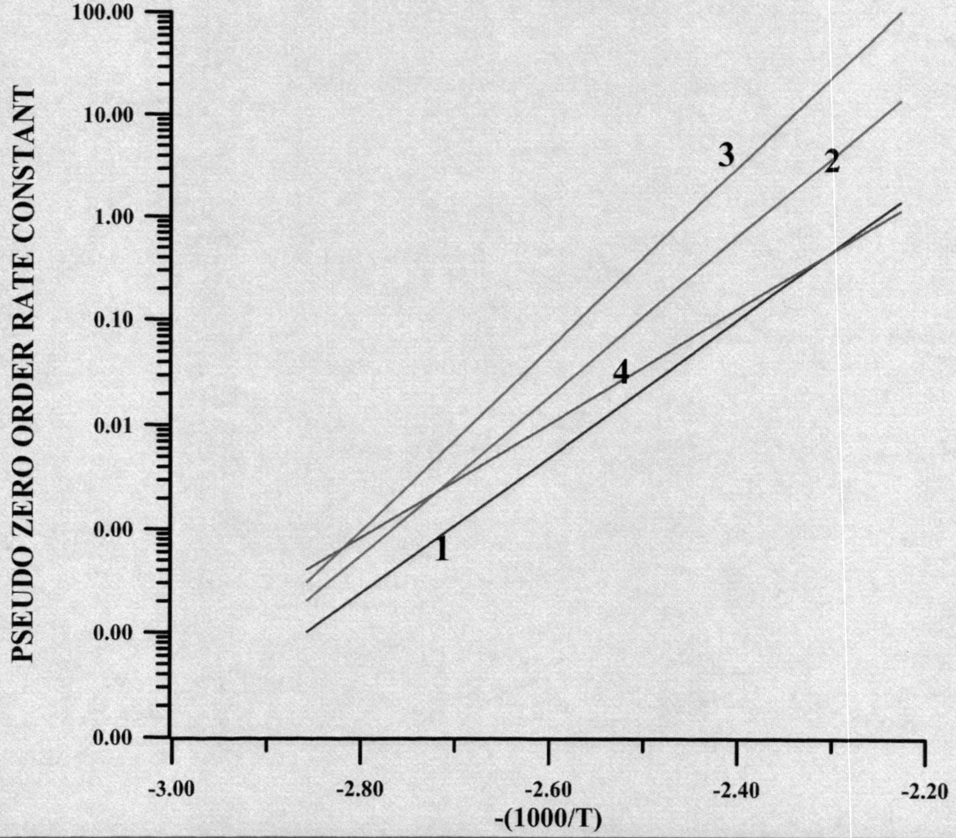

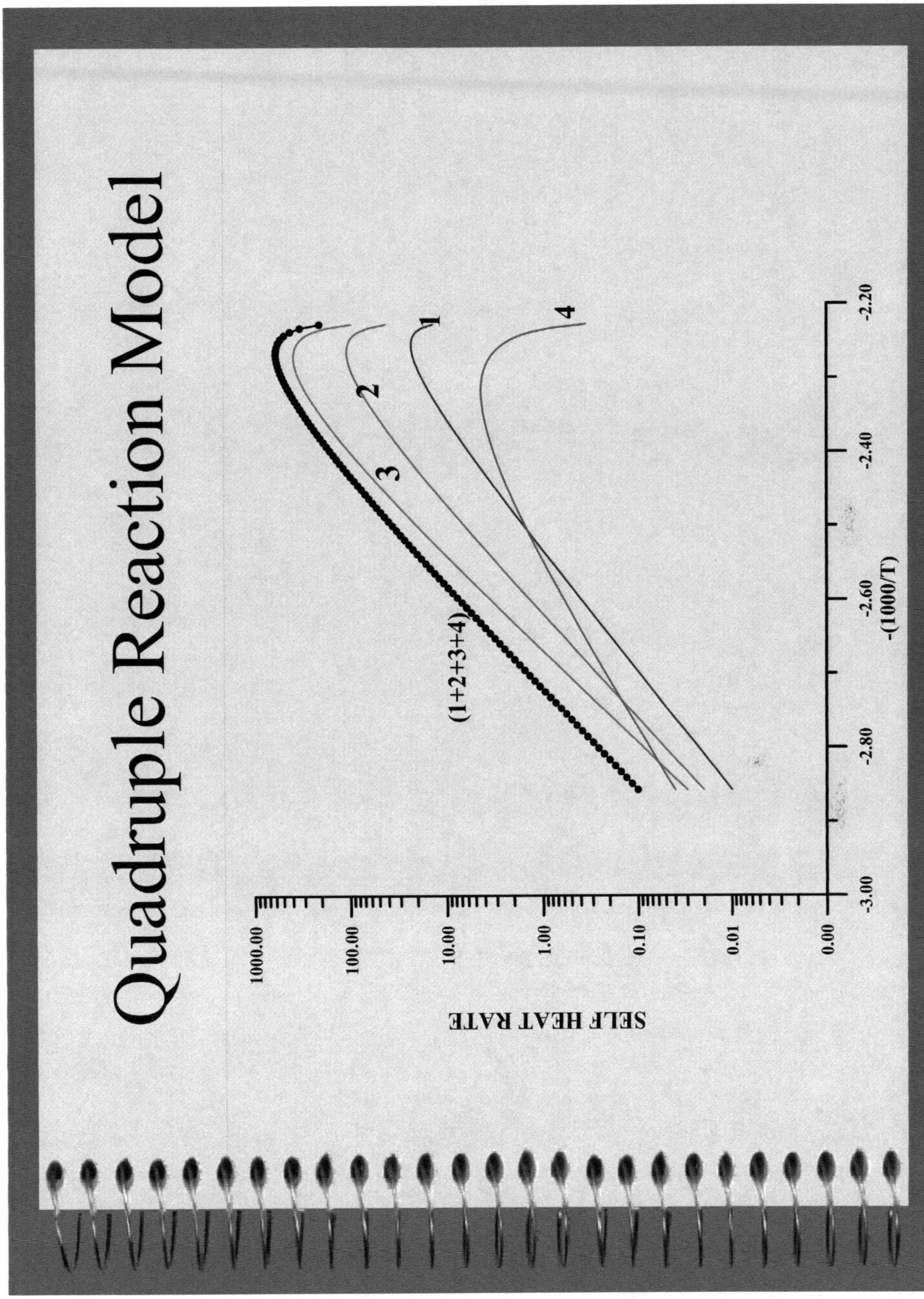
Quadruple Reaction Model
SELF HEAT RATE
1000.00
100.00
10.00
1.00
0.10
0.01
0.00
-3.00
-2.80
-2.60
-2.40
-2.20
-(1000/T)
(1+2+3+4)
1
2
3
4

Multiple Reaction Modeling Approach

Assumption

The heat liberated during the course of an exothermic reaction (as measured by an adiabatic calorimeter) is the cumulative effect due to one or more singular reactions, which could be either exothermic or endothermic in nature .

Model Representation

$$m_T = \sum_{i=1}^{N} m_{T\,i} = \sum_{i=1}^{N} m_{o\,i} \left(\frac{T_f - T}{T_f - T_o} \right)^{n_i} e^{\frac{E_i}{R}\left(\frac{1}{T_o} - \frac{1}{T} \right)}$$

Where,

$m_{T,i}$: self-heat rate corresponding to each reaction

(Positive if exothermic and negative if endothermic),

N : number of reactions occurring in the multiple reaction system,

$m_{o,i}$: initial self-heat rate for i^{th} reaction,

T_o : runaway reaction onset temperature,

T_f : runaway reaction final temperature,

n_i : order of ith reaction,

E_i : activation energy for the ith reaction.

Algorithm

(1) Read adiabatic calorimeter data consisting of time, temperature and self-heat rate.

(2) Assume for i = 1 to N,

$$m_T = m_{T_1} + m_{T_2} + \ldots + m_{T_i} \qquad \ldots(1)$$

Where,

$$m_{T\,i} = m_{oi}\left(\frac{T_f - T}{T_f - T_o}\right)^{n_i} e^{\frac{E_i}{R}\left(\frac{1}{T_o}-\frac{1}{T}\right)} \qquad \ldots(2)$$

(3) Rearranging Eq. (1) we get,

$$m_T = m_{T1}\left(1 + \frac{m_{T_2}}{m_{T_1}} + \ldots + \frac{m_{T_i}}{m_{T_1}}\right) \qquad \textbf{...(3)}$$

(4) From Eqs. (2) and (3) it follows that,

$$\frac{m_T}{\left(1+\frac{m_{T2}}{m_{T_1}}+\ldots+\frac{m_{Ti}}{m_{T_1}}\right)} = m_{o_1}\left(\frac{T_f - T}{T_f - T_o}\right)^{n_1} e^{\frac{E_1}{R}\left(\frac{1}{T_o}-\frac{1}{T}\right)} \qquad \textbf{...(4)}$$

(5) Taking logarithm on both sides of Eq.(4),

$$\log\left(\frac{m_T}{\left(1+\frac{m_{T2}}{m_{T1}}+\ldots+\frac{m_{Ti}}{m_{T1}}\right)}\right)=\log\left(m_{o_1}\right)+n_1\log\left(\frac{T_f-T}{T_f-T_o}\right)+\frac{E_1}{R}\left(\frac{1}{T_o}-\frac{1}{T}\right) \qquad \textbf{...(5)}$$

(6) At the maximum self-heat rate,

$$\frac{d^2T}{dt^2}=0$$

$$nRT_m^2+ET_m-ET_f=0$$

$$E=\frac{nRT_m^2}{\left(T_f-T_m\right)} \qquad \textbf{...(6)}$$

(7) Substituting Eq.(6) in Eq. (5) and rearranging,

$$\log\left(\frac{m_T}{1+\frac{m_{T_2}}{m_{T_1}}+\ldots+\frac{m_{T_i}}{m_{T_1}}}\right)=\log(m_{o_1})+n_1\left[\ln\left(\frac{T_f-T}{T_f-T_o}\right)+\frac{T_m^2}{\left(T_f-T_m\right)}\left(\frac{1}{T_o}-\frac{1}{T}\right)\right] \quad \textbf{...(7)}$$

(8) Solving Eq. (7) using linear regression method and assuming initially that $m_{Ti} = 0$, we get m_{o1}, n_1 and E_1.

(9) Evaluate m_{T1} using Eq. (2) and set $m_T = m_T - m_{T1}$.

(10) Repeat steps (3) to (9) assuming that
$m_T = m_{T2} + \ldots + m_{Ti}$ and then evaluate m_{T2}.

(11) Iterate till results converge.

(12) Print model parameters for all N multiple reactions (N =1 for singular reaction).

(13) STOP

Multiple Reaction Model Prediction for Di-Tertiary Butyl Peroxide System

ARC Data for DTBP System

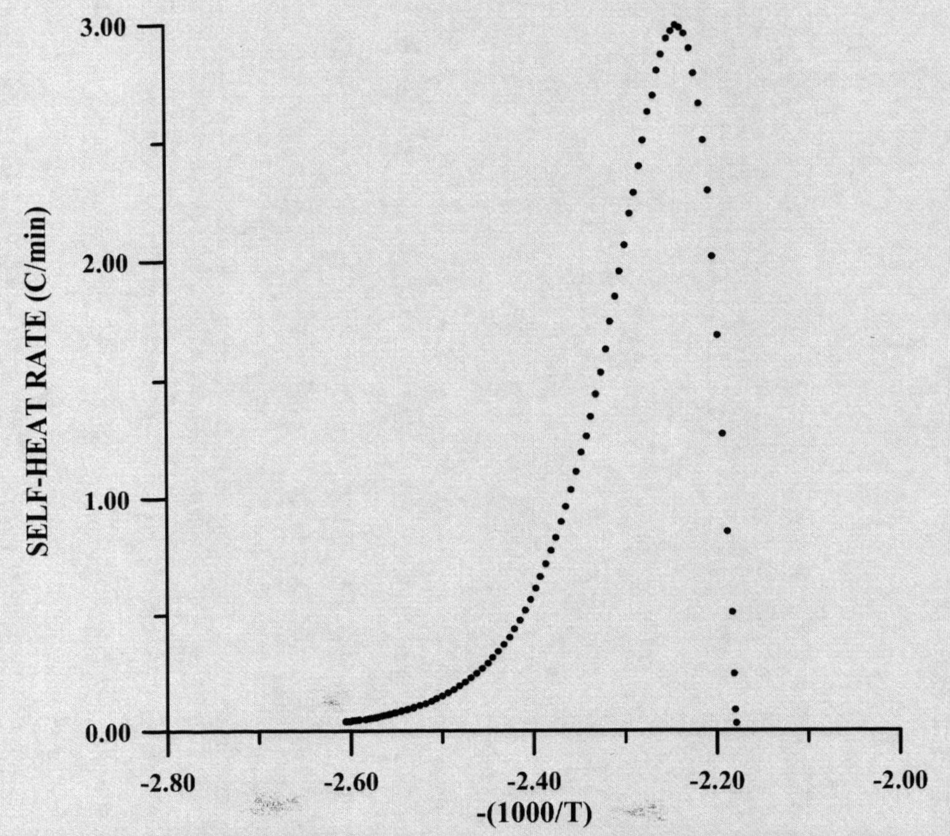

Singular Reaction Model Prediction

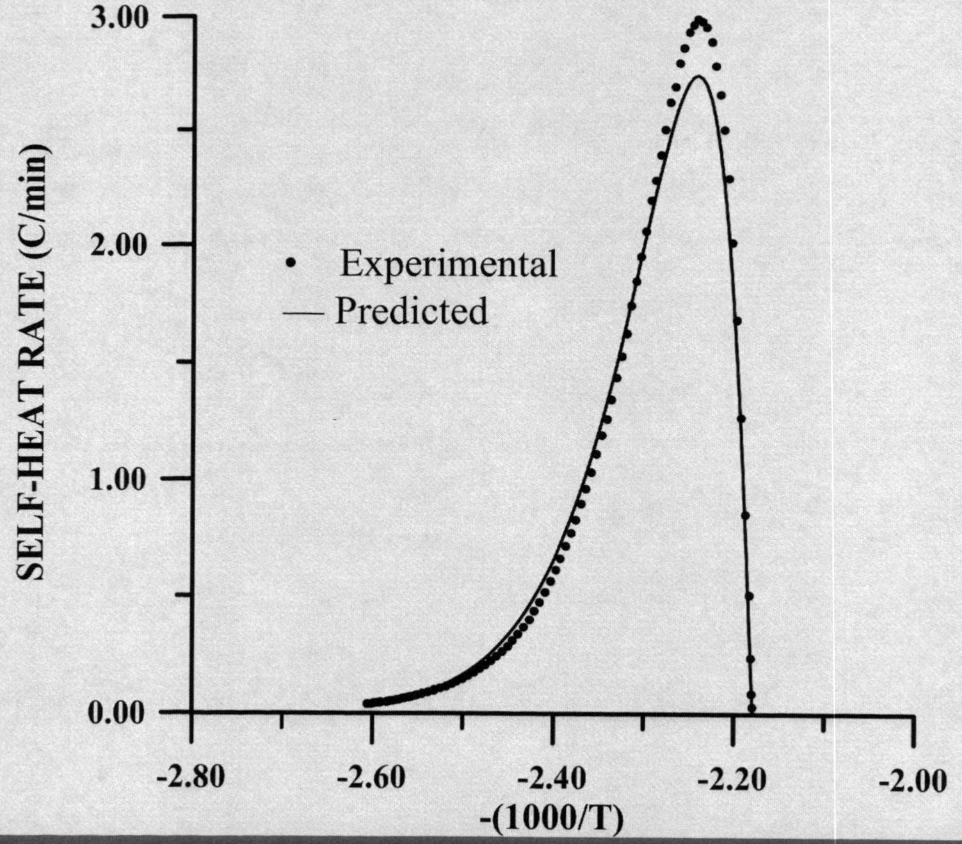

Model Equation

$$m_T = 0.034\left(\frac{459-T}{459-393}\right)^{1.13} e^{\frac{34,899}{1.987}\left(\frac{1}{393}-\frac{1}{T}\right)}$$

Double Reaction Model Prediction

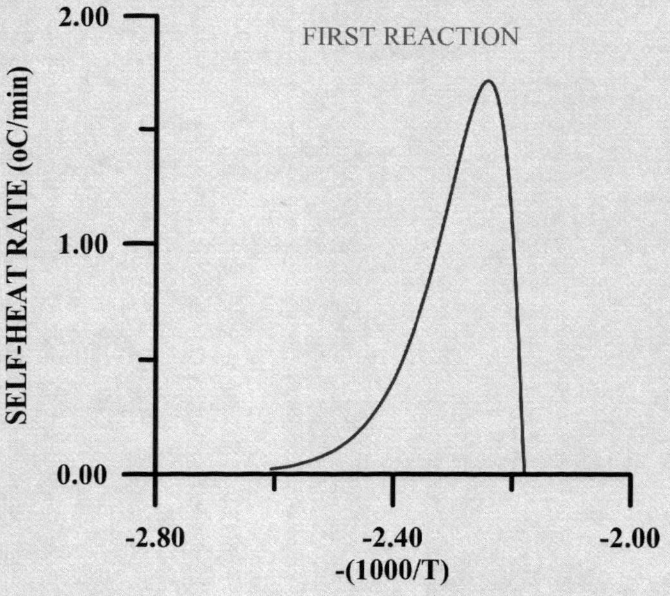

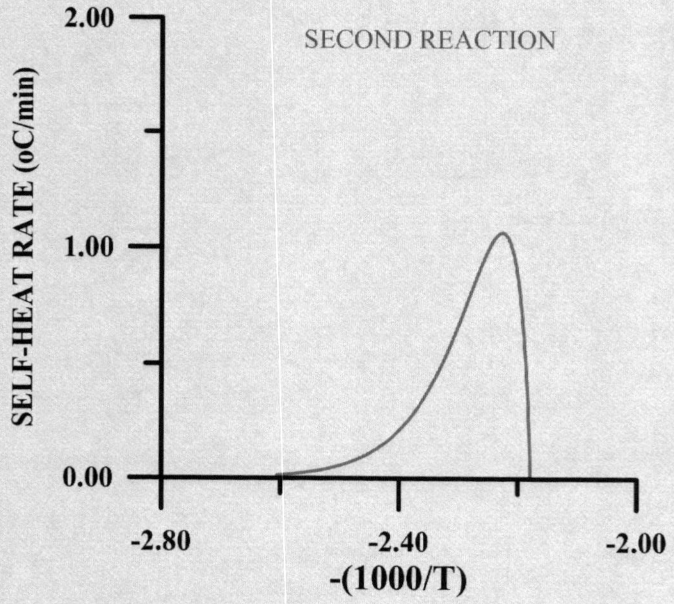

Contd…

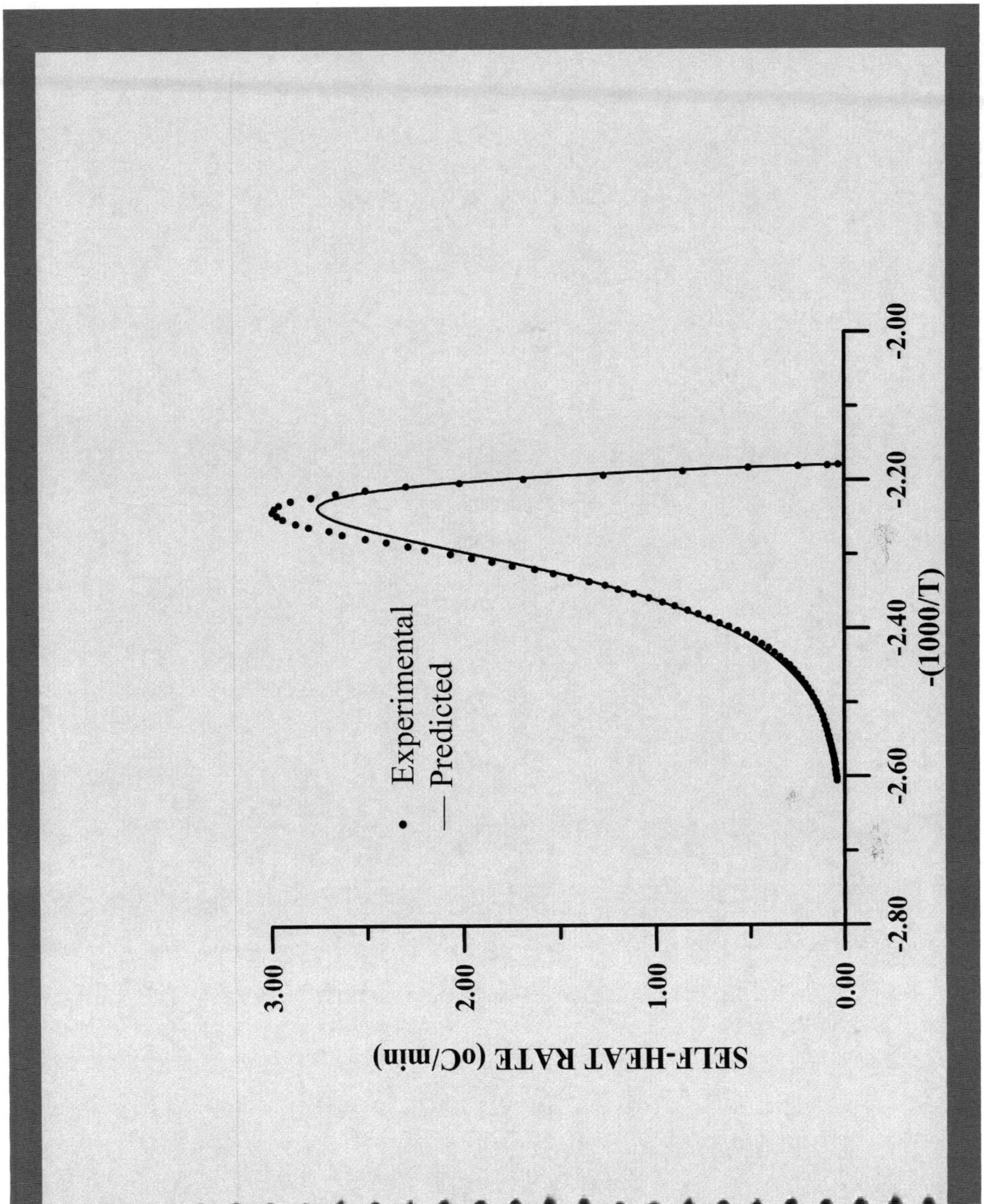
SELF-HEAT RATE (oC/min)
-(1000/T)
• Experimental
— Predicted
0.00
1.00
2.00
3.00
-2.80
-2.60
-2.40
-2.20
-2.00

Model Equation

$$m_T = 0.0210\left(\frac{459-T}{459-383}\right)^{1.1} e^{\frac{34,836}{1.987}\left(\frac{1}{383}-\frac{1}{T}\right)}$$

$$+0.0127\left(\frac{459-T}{459-383}\right)^{0.8} e^{\frac{31,637}{1.987}\left(\frac{1}{383}-\frac{1}{T}\right)}$$

Triple Reaction Model Prediction

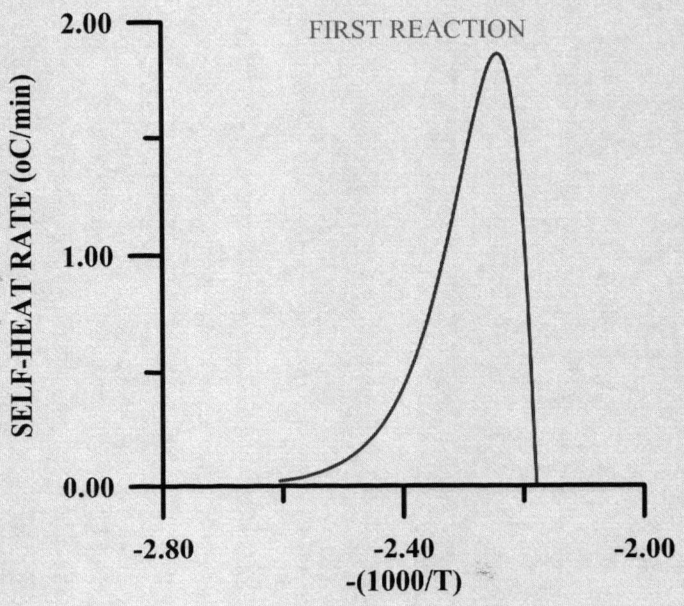

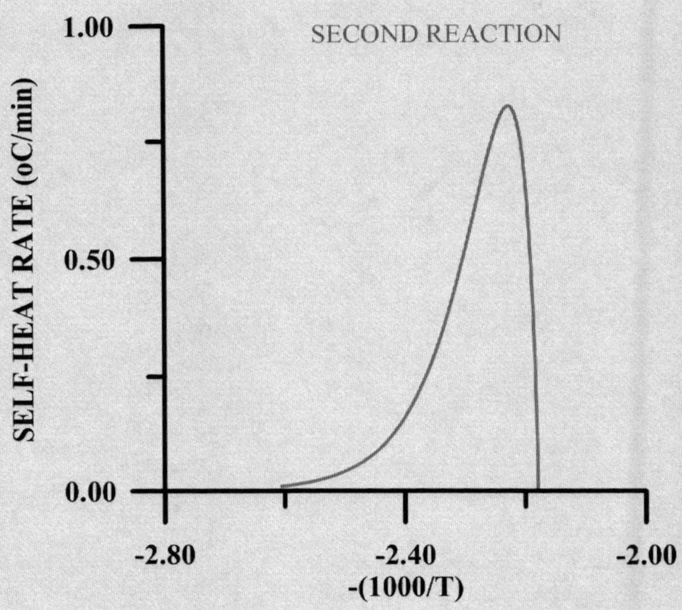

Contd…

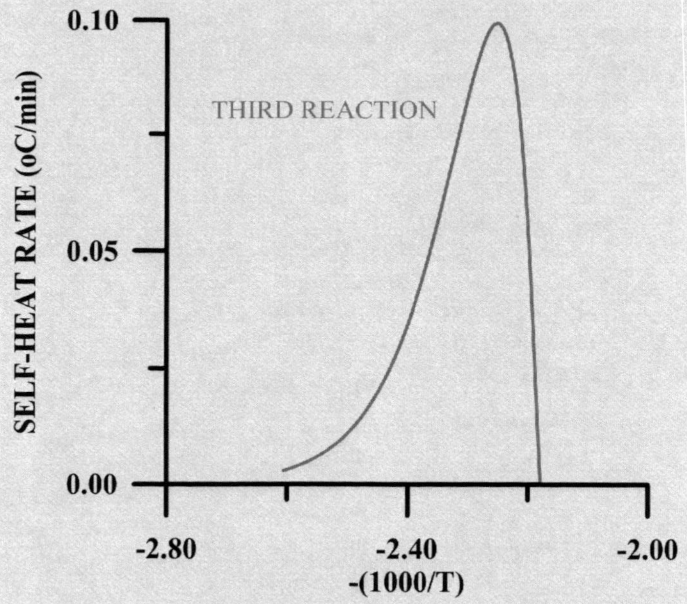
0.10
0.05
0.00
SELF-HEAT RATE (oC/min)
THIRD REACTION
-2.80
-2.40
-2.00
-(1000/T)

Contd...

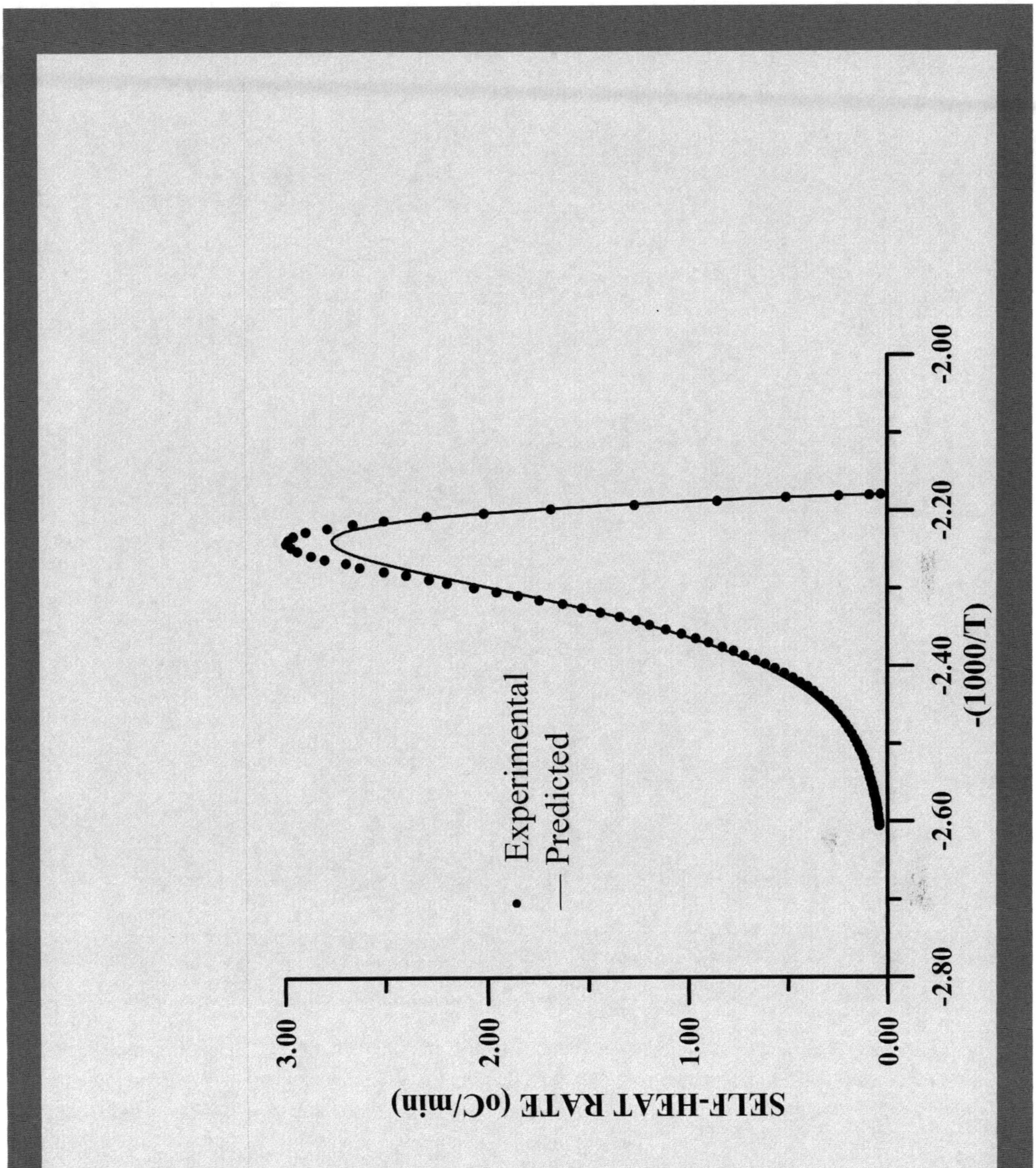

SELF-HEAT RATE (oC/min)
3.00
2.00
1.00
0.00
-2.80
-2.60
-2.40
-2.20
-2.00
-(1000/T)
• Experimental
— Predicted

Model Equation

$$m_T = 0.021\left(\frac{459-T}{459-383}\right)^{1.2} e^{\frac{35{,}585}{1.987}\left(\frac{1}{383}-\frac{1}{T}\right)}$$

$$+0.01\left(\frac{459-T}{459-383}\right)^{0.8} e^{\frac{31{,}540}{1.987}\left(\frac{1}{383}-\frac{1}{T}\right)}$$

$$+0.0028\left(\frac{459-T}{459-383}\right)^{1.1} e^{\frac{30{,}195}{1.987}\left(\frac{1}{383}-\frac{1}{T}\right)}$$

Quadruple Reaction Model Prediction

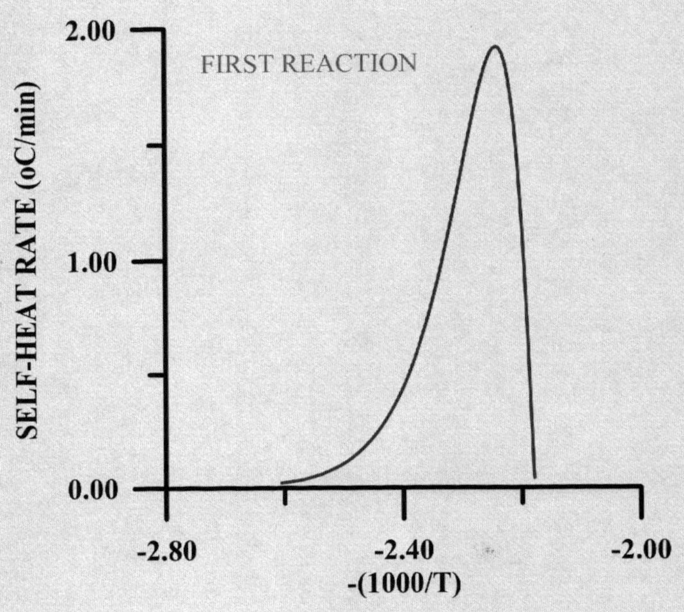

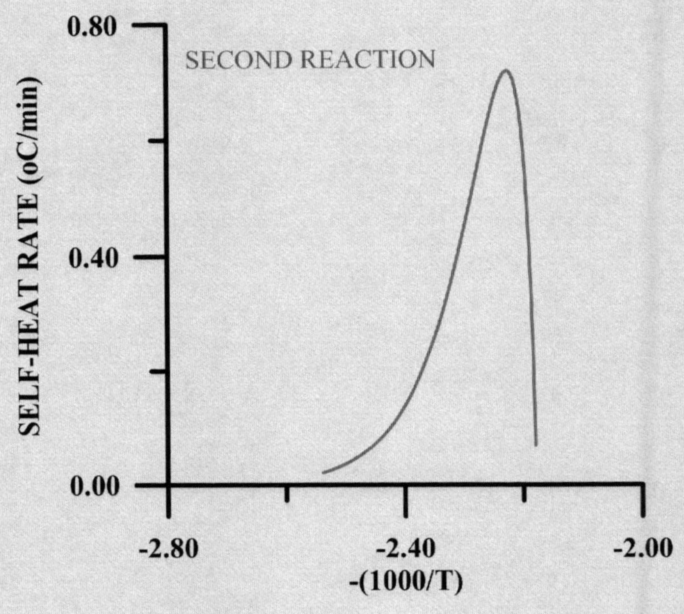

Contd…

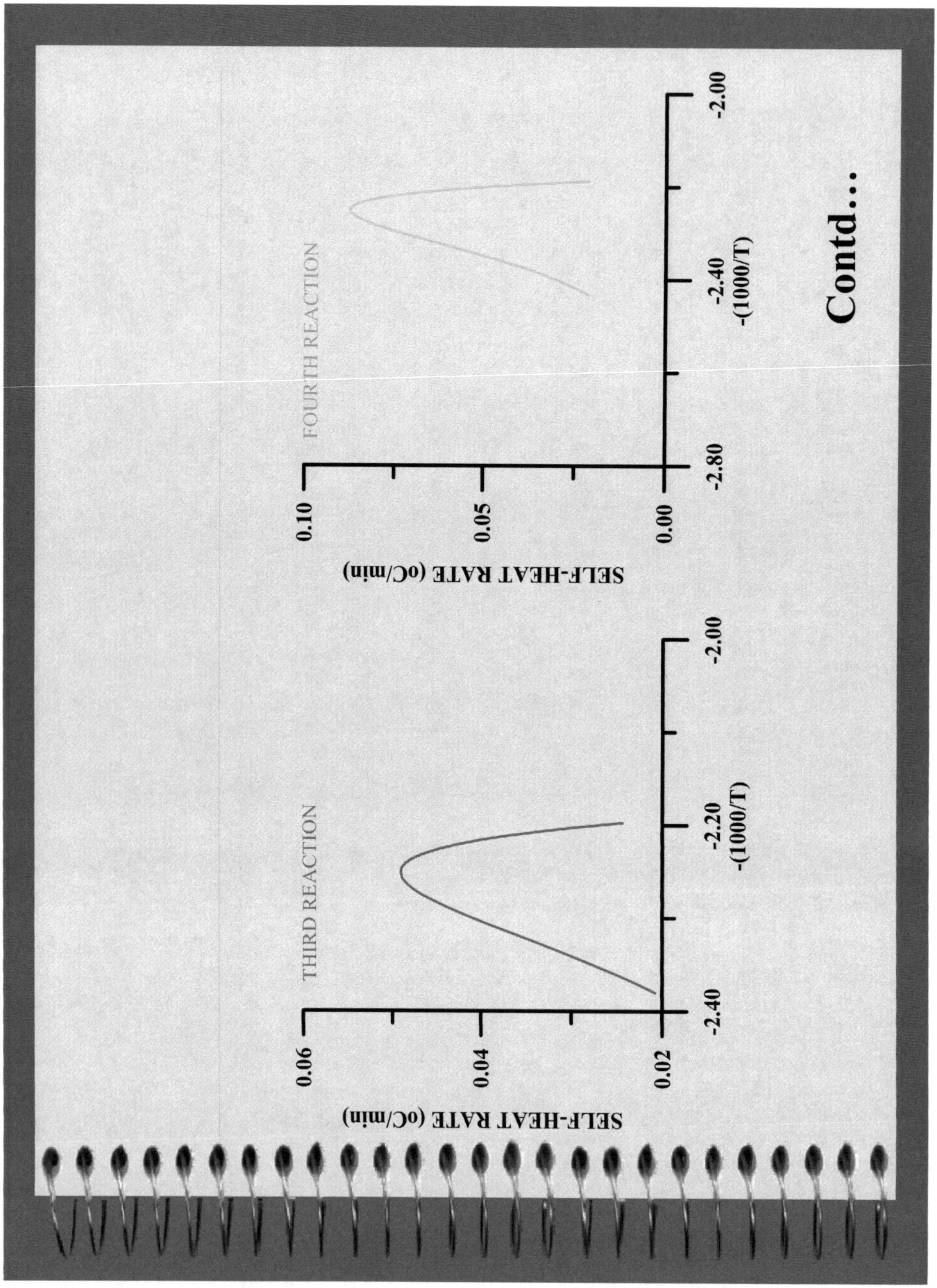
THIRD REACTION
SELF-HEAT RATE (oC/min)
0.06
0.04
0.02
-2.40
-2.20
-2.00
-(1000/T)
FOURTH REACTION
SELF-HEAT RATE (oC/min)
0.10
0.05
0.00
-2.80
-2.40
-2.00
-(1000/T)
Contd...

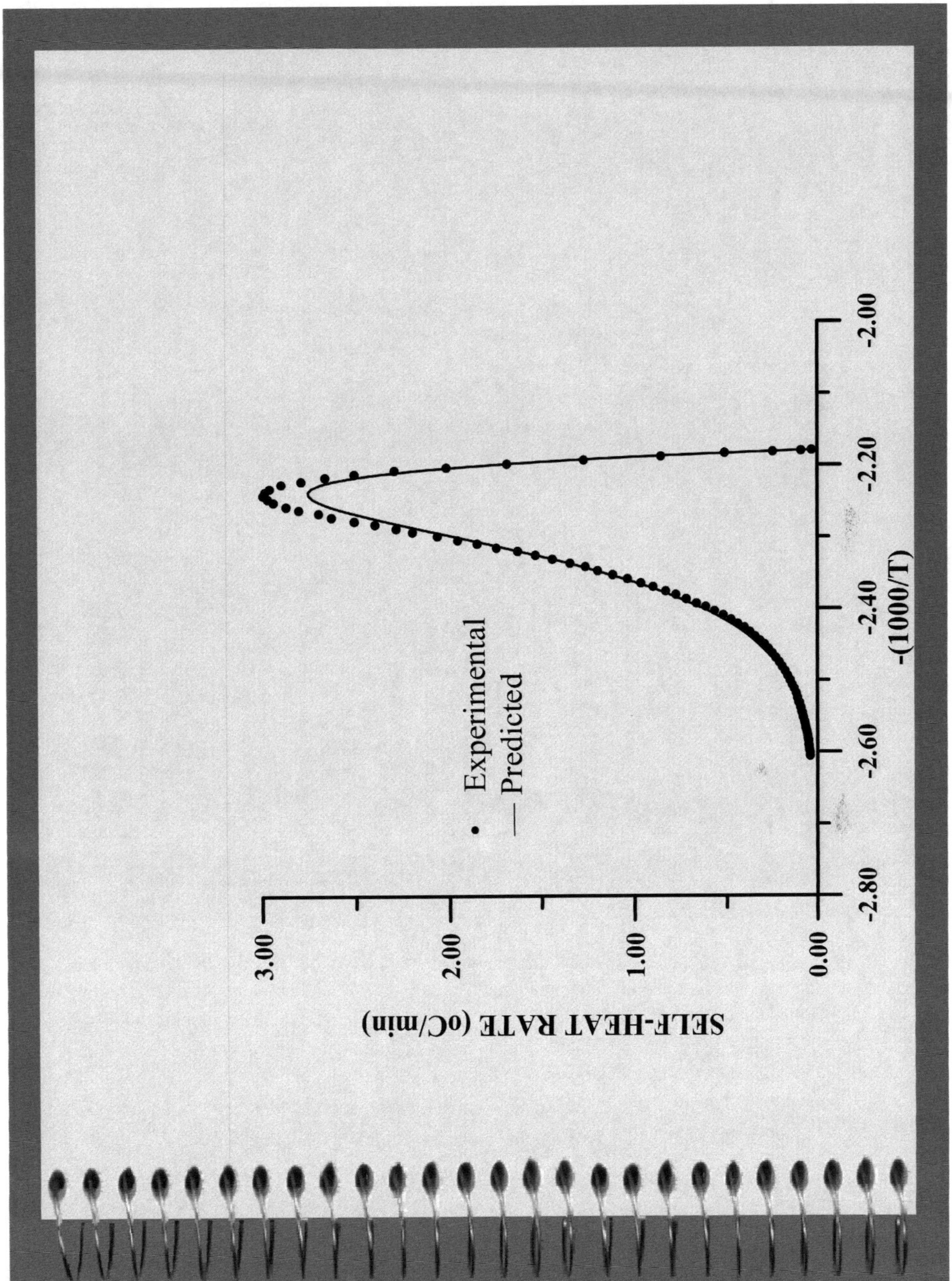
SELF-HEAT RATE (oC/min)
3.00
2.00
1.00
0.00
-2.80
-2.60
-2.40
-2.20
-2.00
-(1000/T)
• Experimental
— Predicted

Model Equation

$$m_T = 0.021\left(\frac{459-T}{459-383}\right)^{1.2} e^{\frac{36{,}029}{1.987}\left(\frac{1}{383}-\frac{1}{T}\right)}$$

$$+0.0078\left(\frac{459-T}{459-383}\right)^{0.8} e^{\frac{31{,}747}{1.987}\left(\frac{1}{383}-\frac{1}{T}\right)}$$

$$+0.0018\left(\frac{459-T}{459-383}\right)^{1.1} e^{\frac{30{,}308}{1.987}\left(\frac{1}{383}-\frac{1}{T}\right)}$$

$$+0.0023\left(\frac{459-T}{459-383}\right)^{1.1} e^{\frac{29{,}993}{1.987}\left(\frac{1}{383}-\frac{1}{T}\right)}$$

Applications

Reaction Mechanistic Studies

FIRST REACTION

DTBP → 2 ACETONE + ETHANE

SECOND REACTION

DTBP → ACETONE + METHYL ETHYL KETONE + METHANE

THIRD REACTION

DTBP → ACETONE + ISO-BUTYLENE OXIDE + METHANE

FOURTH REACTION

DTBP → ISO-BUTYLENE OXIDE + t- BUTANOL

OVERALL REACTION

4 DTBP → 4 ACETONE + 2 IBO + t-Bu-OH + MEK + 2 METHANE + ETHANE

Thermal Inertia Factor (Phi) Correction

$$\frac{1}{T_{on_{ad_i}}} = \frac{1}{T_{on_i}} + \frac{R}{E_i} \ln \varphi$$

$$T_{ad_i} = T_{on_{ad_i}} + \varphi \cdot (T - T_{on_i})$$

$$m_{T_{ad_i}} = \varphi \cdot m_{T_i} \exp\left[\frac{E_i}{R} \left(\frac{1}{T} - \frac{1}{T_{ad_i}} \right) \right]$$

$$m_T = \sum_{i=1}^{N} m_{T_{ad_i}}$$

Vent Sizing Evaluation

Heat generation rate:

$$q(\text{kW/kg}) = C_p \cdot m_T$$

Reaction Inhibition

High pad gas pressure inhibits formation of DTBP vapors, resulting in reduced production of noncondensibles like ethane and methane i.e., inhibits first, second and third reaction.

Quench Tank and Dump Tank Design

Evaluation depends on the corrected values for maximum temperature and average heat generation rate during runaway reaction.

Thank you!

Planning of an Olefins and Aromatics Plant

A. Fraga[1], E. Terra[1], Arturo Cervantes[2], and Bhieng Tjoa[2]. (1) BRASKEM, Salvador, Brazil, (2) Optience Corporation, Campbell, CA 95008

The petrochemical market is more competitive and global than ever. The raw materials and products can come from any place in the world. Associated to this, the geopolitical tensions and the speculation make the world-wide market of oil extremely volatile, directly influencing the petrochemical market, which often does not follow this speculation.

On the eve of another "low cycle" in the petrochemical market, because of projects with large capacities and high technology, the result is a higher offer than demand and a reduction in the profitability of the companies.

Competitiveness is the key factor for the companies that are preparing themselves for these challenges, among other actions, through an efficient choice of its raw materials, the best use of its assets and the "exploration" of the best markets.

The volatility of the prices, sometimes favoring the olefins, sometimes the aromatics products and the complexity and diversity of Braskem's processes which include Reforming and Cracking units, strengthen the necessity of searching the best point between raw materials, market and operation.

To solve this highly complex problem, the acquisition of an advanced optimization application, gives these companies a competitive advantage. One year and half ago Braskem acquired Optience SCMart Suite as the optimization application platform to help their planning and scheduling process.

The main model, developed in SCMart, is a multiperiod nonlinear planning model which includes all plant constraints and uses a low order nonlinear polynomial correlation yield model based on SPYRO results for the cracking furnaces. It also includes a second order polynomial to estimate the reformer yields as well as a complex model for a naphtha fractionator. The planning model is supported by a scheduling model to achieve optimal results at the daily level. The system also uses a feedstock database library that tracks the different available feedstock qualities and properties.

In this work, we discuss the planning module that is used to support the technical and economical decisions at Braskem. The nonlinear planning model consists of about 5,000 variables per period including flows, inventories, properties and other process variables. It covers two ethylene plants, one reformer plant, plus other C4s, C5s and aromatics plants, in which the main products are ethylene, propylene, butadiene, isoprene, benzene, para-xylene and other aromatics. These process variables include COTs, reforming and cut temperatures, possible co-cracks, furnace and reformer utilization, ethane recycling flows and loads to other downstream units.

We present different case studies and scenarios which have been run during these last months that demonstrate how the tool has been helping us in the decisions. The study of these scenarios has been helping us to decide the amount and quality of the Naphtha to be bought and processed in each unit. It has also helped us to identify the ideal markets and in the exploration of trade offs and ideal yields, as well with the optimal use of the assets, through concepts of shadow price and profit variability.

Applying Attainable Regions to Scheduling

David Glasser[1], Diane Hildebrandt[1], Tumisang Seodigang[1] and Imtiaz Laher[2]

(1) Centre of Material and Process Synthesis (COMPS), School of Chemical and Metallurgical Engineering, University of the Witwatersrand, Johannesburg, 1 Jan Smuts Avenue, Johannesburg, South Africa,
(2) School of Chemical and Metallurgical Engineering, University of the Witwatersrand, Johannesburg, 1 Jan Smuts Avenue, Johannesburg, South Africa

The Attainable Region (AR) has been used to optimize the structure of steady flow reactors. Horn (1964) first defined the Attainable region. Horn noted the AR represented the full totality of all physically possible reactors. Glasser et al. (1987) developed the geometric properties, as well as proposed a constructive approach for determining such a region. Glasser et al. furthermore defined the necessary conditions for any candidate AR. The AR research has continued in the field of steady flow reactors. More recently Seodigeng (2006) noted the AR analysis method can be used to identify optimal reactor types, flow configurations and key design parameters which will optimise the objective function. This paper attempts to produce an analogy of an AR in batch reactors with the production rate as the defining objective function.

The approach taken in this paper is purely geometric/graphical at present and does not follow the approach as proposed and highlighted by Bryson et al. (1975) which uses a purely mathematical approaches to optimise a said objective function. In this case this said function would be production rate. Note however these methods are able to deal with much higher complexities.

This paper focuses on the group of reaction kinetics that causes concavity in the batch reactor profile. Such kinetics are typified by a maximum in the rate for any reactant. Typical forms of such kinetics are the autocatalytic kinetics, or the more advanced reaction rates with deactivation terms, as is found in enzyme kinetics with parallel deactivation (San et. al, 1983). The concavity caused by these kinetic forms allows us to relate the maximum production rate to features on the batch profile, when drawn in the concentration and time space ([ ]-x-axis, t-y-axis). For the purpose of this discussion the production rate will be defined as mol/time, note the increases in this will lead to equivalent increases in the volume/time (for a given product concentration) space. For simplicity filling and emptying will be regarded as instantaneous (i.e. time is small when compared to reaction time), with a perfect mixing assumption.

From the definition of production as mol/time, the batch production rate can be related to the inverse of the gradient of a line emanating from the feed conditions to the product conditions multiplied by a factor of $\pm aV$, where $\boldsymbol{a}$ represents the stoichiometric coefficient, V the volume and the positive or negative will relate to whether the product or reactant space is being considered. The lowest gradient was found to lead to the greatest production rate. Thus a line from the feed conditions to the lowest tangential point on the batch profile will lead to the maximum production rate possible in a standard batch. This maximum production rate is only possible at one exit concentration. By a simple graphical approach this point can be used to effectively determine further increases in production by simple control changes such as partial emptying of batches or more advanced control changes of converting to a Fed Batch reactor, for all exit concentrations. Furthermore this approach allows us to determine optimal points, as well as regions where the production will be greater but not necessarily optimal.

The production rate increases taken over an infinite series of batches or a Fed Batch reactor with beginning infinitely small volumental increases can be related to steady flow reactor systems. Thus a relationship between the optimal structure of a steady flow system and the optimal control and operating policy of a batch system will be shown. In effect the production rate by implementation of this technique will asymptote to values related to steady flow reactor systems. The batch fed option is similar to that explained by Glasser et. al. (1985) using Zwietering's maximum-mixed reactor model. This model analysed a tubular reactor with feed streams entering along the length. At all stages in the reactor the material was well mixed across the cross-section but no mixing occurred longitudinally (Glasser et. al, 1985). The model of a PFR shown by Glasser et. al (1985) shows the feed of the side-streams decreasing geometrically away from the exit, the results found here reflect similar geometric feed rate results. These results are illustrated using examples.

References:

Bryson, A. E. and Ho, Y. C., 1975, *Applied optimal control*, Washington DC: Hemisphere.

Glasser D., Crowe C.M. and Jackson R., 1985, *Zwietering's Maximum-Mixed Reactor Model and the Existence of Multiple Steady States*, Chemical Engineering Communication, Vol. 40, pgs. 41 to 48

Glasser D., Crowe C.M. and Hildebrandt D., 1987, *A Geometric Approach to Steady Flow Reactors: The Attainable Region and Optimization in Concentration Space*, Ind. Eng. Chem. Res., Vol 26(9), 1803

Horn, F., 1964, *Attainable and Non-Attainable Regions in Chemical Reaction Technique*, Third European Symposium on Chemical Reaction Engineering,
Pergamon, New York, pp. 1-10.

Seodigeng T.G., 2006, *Numerical Formulations for Attainable Region Analysis*, Ph.D. Thesis, University of the Witwatersrand, pg. 12

Keeping Advanced Process Control (APC) Performance and Maintenance and Improvement of APC System

Satoshi Ooyama, Mitsubishi Chemical Corporation, Japan
Koichi Onodera, Mitsubishi Chemical Corporation, Japan
Soichi Amano, Mitsubishi Chemical Corporation, Japan
Minoru Yoneda, Mitsubishi Chemical Corporation, Japan

Abstract

This paper explains the history of advanced control projects in Mitsubishi Chemical Corporation (MCC) and the current problems of them. Our company started using advanced control systems in the 1980's which were made by MCC. Afterwards, we changed them gradually into commercially packaged software (mainly multivariate model prediction controllers), like DMC plus for large-scale plants. Using the commercial packages helped us to obtain greater profits much faster than before. Most large-scale petrochemical plants already have been equipped with the advanced control systems, and we have now shifted the stage of the projects to development of systems for small and medium-sized scale plants. Also we recognized the importance of system maintenance to keep the advanced process control (APC) performance high, and daily maintenance and improvement of control performances have been done. Simultaneously, we have been trying to use a couple of new technologies like soft sensors to improve control performance. We have also been investigating methods of quantitative assessments of control performance to detect deterioration much faster and the method of efficient model construction to reduce costs and time to apply them to the new plants.

1. Introduction

APC has been performing an important role in Mitsubishi Chemical Company in these 10～20 years to conserve energy, to increase output and to reduce operation loads of more than 20 plants. Now our APC projects concentrate not only on new developments of the systems but also the maintenance of them. This paper discusses the history of APC and the current and future problems of them.

2. A History of Advanced Control in Mitsubishi Chemical Corporation

The word, APC, mentioned here includes not only multivariate model prediction control but also conventional advanced control based on PID. However, a simple cascade control, the ratio control, and the sequence control were excluded.

First Phase ('80s)

The spread use of DCS and process computers enabled us to apply advanced control programs.

Our applications were LQ and LQI control and the fuzzy control etc. At that time, the APC software was coded by our engineers and therefore it was technically unique and advanced but insufficient for maintainability.

As a result, the long duration of software development sometimes resulted in poor cost performance, or poor maintainability prevented their continuous use.

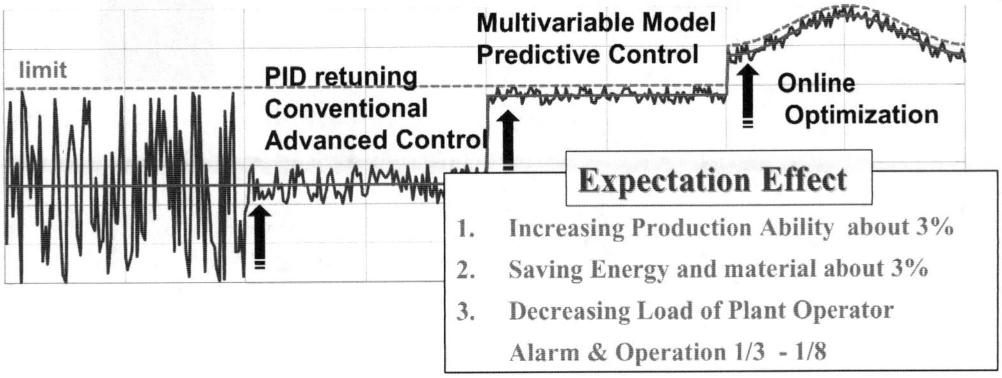

Results: Plant number of Applied of APC

Retuning Conventional Control	Conventional Advanced Control	Multivariable Model Predictive Control	On line Optimization
75	59	36	4

Figure 1. The Purpose and the application number of APC

Second Phase ('90s)

Process computer's performance became high and the use of computer networks became common at that time. This made the APC developments much easier than before and a lots of commercial packages of advanced control were released by domestic and overseas vendors, which reported a lot of success by using them. We also applied one of the packages to a couple of large plants such as ethylene, and obtained increased profits by APC.

Prompt development of software was most important then, and we used the support of the software vendors to make implementation faster and to obtain profits earlier.

Third Phase ('00s)

As most of the large-scale plants had already been equipped with APC system by 2000, we focused on the development of APCs for medium-sized plants. We developed a small scale APC system by ourselves because it is not cost efficient to invest in software for commercial small and medium sized plants.

At the same time, maintenance of the current APC systems became important because dynamic models for APC must be updated when plants are changed to increase throughput or some other modification. Also, we have been investigating new functions such as soft sensors and operation guidance systems to improve the performance of APC systems. New applications of APC have also been considered such as APC for plant emergency support.

3. Importance of maintenance of Advanced Control System

Our company has been using model predictive control (MPC) systems like DMC since the 90's and the effect of MPC (ability increase, conservation of energy, and reduction in the number of operators) adds up to more than $10 million. However, remodeling of the plant at each shutdown maintenance (once a year, every two or four years) sometimes causes model mismatches and the performance of MPC deteriorates in those cases. The importance of the maintenance of software is not understood so widely as the necessity of maintenance of the hardware. However, the APC model mismatch caused by changes in plant result in the deterioration of control performance, which is actually shown as the decrease of the time APC is on use, deterioration of energy savings or operator loads.

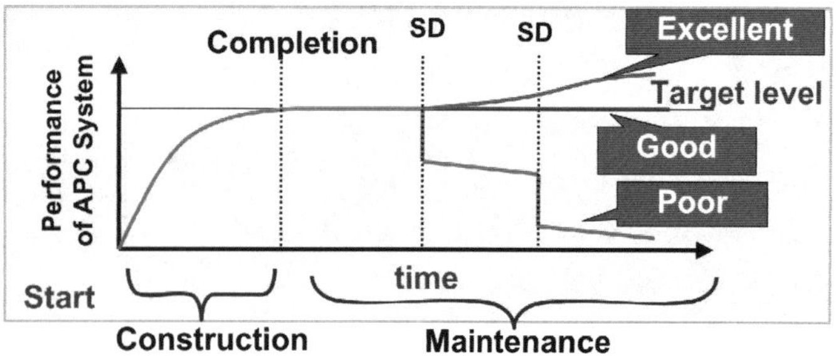

Figure 2. Importance of APC System Maintenance

Figure 2 shows the relation between the profits of the APC systems(vertical) and the duration from the first introduction of APC systems (horizontal).

If the MPC model is properly maintained, we also maintain profits at the proper level. On the other hand, without the model modification at the time of the plant change, APC may perform very poorly and profits will go down drastically. In the worst situation, APC itself may be given up. We would like to not only keep the model performance at the initial level but also to improve it so as to dramatically increase the profits.

Figure 3 is a turning-on rate of a plant in MCC. The turning-on rate has fallen down to 20% here. In this plant, plant modifications were so often that remodeling of MPC could not be kept up with. This mismatch resulted in the deterioration of control performance and degradation of the turning-on ratio. Loss caused by the poor performance added up to a few million which would have been earned by the properly maintained APC. In this case, we clarified the points of plant modifications and designed the efficient remodeling schedule. Less than 3 months' efforts recovered turning on rate to more than 60%.

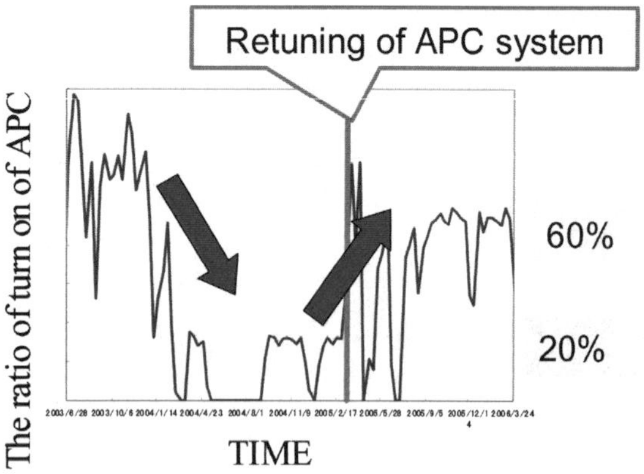

Figure 3. Example of recover turning rate of APC

However, evaluation of the control performance and remodeling is time consuming and effective methods are necessary because limited numbers of engineers can manage the work. Currently we have been examining several control performance evaluation tools and automatic plant test tools with academia for that purpose. Especially for the latter applications, the development of the closed-loop identification tool is promising because we can test the dynamics without disturbing daily operation of the plant.

4. Improvement of APC System

MPC enables multivariate control in the chemical process which has a lot of upper and lower constraints and wide variety of response time.

However, poor performance still remains in sensor systems (chemical composition analyzers, temperature, pressure, flow rate and liquid level, etc.). Recently, to compensate for the inaccuracy and tardiness of sensor systems (especially on-line analyzers), construction of soft sensors became common in many cases. The soft sensor utilizes chemo-metric techniques and has several advantages:

(1) As the sampling period of gas chromatography is so long (around every ten to thirty minutes), it cannot detect sudden disturbances. Soft sensors can interpolate the data between the sampling periods using continuously measured variables like temperature.

(2) Gas chromatography is so complex (it is often called a small chemical plant) that it often stops due to a hardware trouble or indicates wrong measurements, which can make the plant unstable. Soft sensors will be a back-check system of wrong indication of GC because it uses sensor data which are not influenced by GC trouble.

(3) Continuous measurements of several intermediate compositions in the process will be useful for stable operation. In many cases they are measured once a day at most as a measurement for the process capacity index. Soft sensors will be used as a virtual on-line analyzer which gives continuous data estimated by several process values. A resin plant uses soft sensors of melt index which can not be properly measured by any online analyzer. They are more effective for prompt recovery of an unexpected change in melt index than infrequent measurement by off line analyzers.

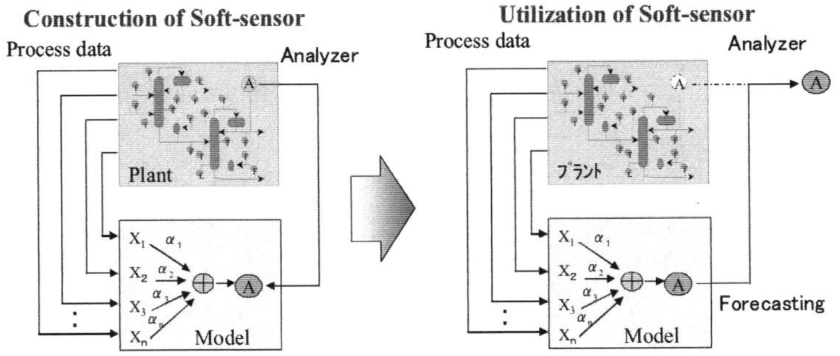

Figure 4. Construction and Utilization of Soft-sensor

Figure 4 is a structure of soft sensor. Using past measurements of the process variables, prediction models are estimated statistically. If these models perform properly, we use them to interpolate infrequent measurements of the analyzer.

Figure 5 is an example of soft sensors which are in use. It predicts the composition changes 0.5 or 2 hours faster than the analyzer and interpolates the GC measurement (once every 5 to 40 minutes). Using the predicted values of the soft sensor for the controlled variable of MPC, control performance is improved dramatically.

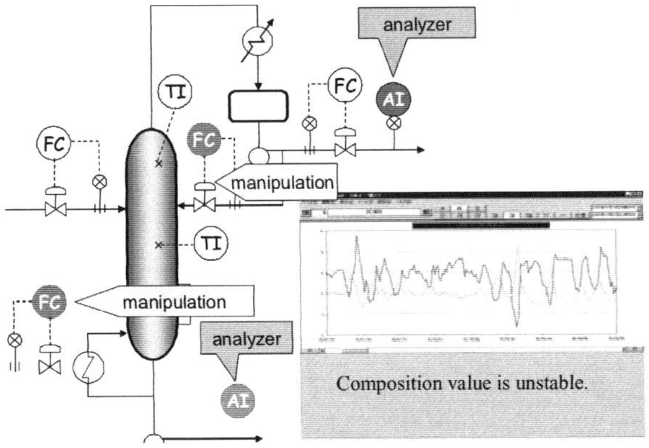

Figure 5-(a). Before Soft Sensor applied

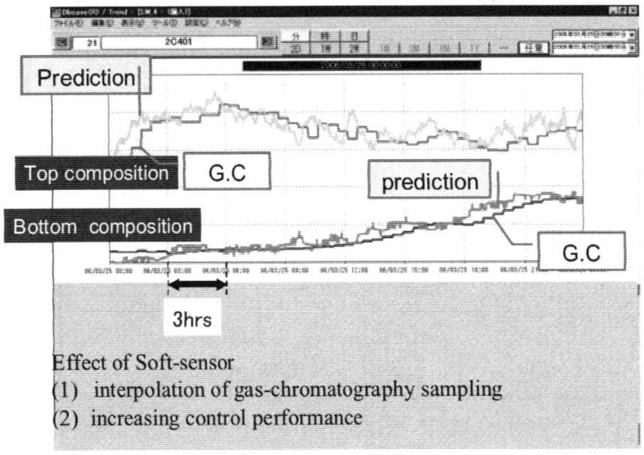

Figure 5-(b). After Soft Sensor applied

In some applications, soft sensors are used for a back-check of important gas-chromatography measurements for quality control. In the past, failure of the fault detection of GCs sometimes caused huge trouble in safety and quality. Helped by the back check systems, such faults are detectable more easily and promptly.

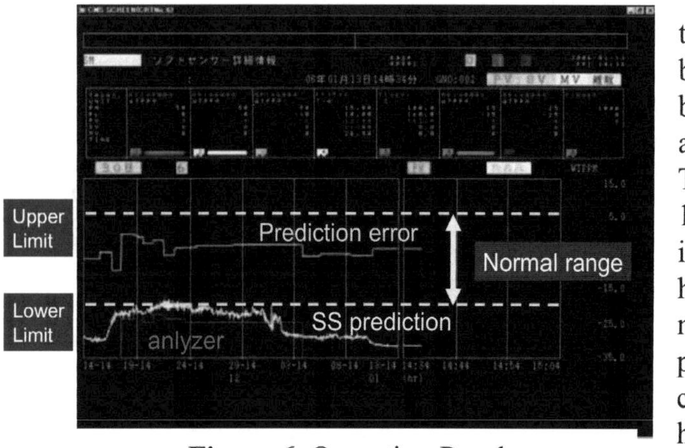

Figure 6. Operation Panel

Figure 6 shows an example of operation displays of the GC measurement back check. When the deviations between measured value and predicted value by SS become greater than a threshold, the system sends an automatic alarm to notify the panel operator.

The number of SS applications goes up to more than 100 in MCC and performs quality and safety improvements effectively. Current problems of SS are how to keep the model performance high. Because SS models are constructed by past operational data, predicted values may be invalid when operational conditions become quite different from the past. We have been trying to develop a maintenance method which detects SS invalidity more easily.

5. The future of APC System

Though APC performs quite important roles in MCC plants, it is not the only factor in safe and effective operations of chemical plants. We would therefore like to discuss operations of chemical plants from a wider viewpoint here. That is: a highly effective operation (conservation of energy and saving resource) and the super-stability operation system (stabilization and absorption of turbulence, etc.).

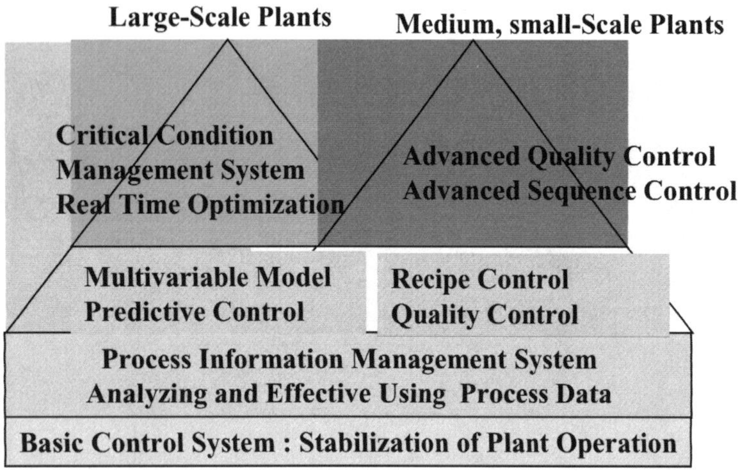

Figure 7. Plant Operation Support System based on APC

Demands for highly effective, super-steady operation systems of chemical plants will increase more and more in the future. To fulfill these needs, APC will perform an important role as an operator support in emergencies or non-stationary stages as start up and shutdown because it can supervise and control the process entirely.

Figure 7 is a target of the operational support systems based on the APC. It will be divided into two types of applications. One is for large-scale petrochemical plants like ethylene, and the other is for small or medium scaled batch plants.

In the former applications, the constraint control operation is advantageous on cost reduction and safety. In other words, when the plants deviate from their constraints, such conditions will be safety accidents or plant shutdown which causes huge loss on the plants.

Therefore, in the future, we are planning to use the APC for abnormal and non-stationary operation support. For instance, at the non-stationary operating conditions like startup or partial shutdown of several units, operators must pay attention to various abnormal phenomena different from normal operations.
Even in such situations, if the multivariate APC manages the non-stationary conditions, the possibilities of abnormal phenomena will be reduced and it will provide good support for panel operators.
Various patterns of non-stationary conditions must be taken into consideration by the APC in these cases. Combinations of logical or sequential control like PROSEG 1) with APC will be effective in these applications.

On the other hand, for the latter applications, the advantage obtained from the constraint control is relatively small and demands and budgets are also quite small for the applications. On the contrary, demands for recipe, quality and production schedule managements are strong for the applications of APC systems. To satisfy these demands,application of the soft sensors or data mining technology etc. will be effective.

We also developped a small sized DCS and MPC package which small scaled plants can afford to use. Using this system, we think that quality control and automatic generation of the control sequence leads to t improvements in productivity.

6. Conclusions

The history and the problem related to our APC system have been described. At this stage we think that the introduction of APC to large-scale plants has finished and scale have advantages already been made due to the best use the APC. From now, the maintenance and improvement of APC performances will be a key issue for a competitive edge. For small sized batch plants, there still remains huge various needs for APC systems. These depend on the characteristics of plants and we have to modify APC to fulfill their various needs.

Efficiencies of APC maintenance are largely dependant on the improvement of the modeling tools and their performance. We will investigate them too.

References

1. R. Batres, S. Lee, M. L. Lu and Y. Naka (1997), "Study on Safety Evaluation and Operational Design for Startup," Process Safety Progress, Vol. 16, No. 1, pp. 37-42

ERROR BAND IDENTIFICATION AND CHARACTERIZATION FOR ADVANCED PROCESS CONTROL (APC) PERFORMANCE ASSESSMENT

Anand Vennavelli and Rob Whiteley, Oklahoma State University, Stillwater, OK
E-mail: anand.vennavelli@okstate.edu, rob.whiteley@okstate.edu

I. Introduction

The term controller performance assessment is used in many ways. The performance assessment of multivariable controllers from a business perspective differs from the control engineer's perspective in its assessment objectives. Engineering analysis is used to identify undesirable dynamic characteristics such as valve stiction, improper controller tuning, and excessive controller action. Controller performance is frequently characterized by comparing the actual process output variance to the output variance of an optimal controller such as the minimum variance controller.

Business analysis, on the other hand, utilizes a longer-term (weeks or months) view [1, 2] of controller performance with a business emphasis on continuous quality improvement, identification of best practices, and allocation of limited resources for control system maintenance. Business analysis methods are implemented using statistical process control (SPC) and "six-sigma" principles [3].

SPC analysis was originally confined to the parts industry until the "quality revolution" triggered its application in the process industries as well [4]. Much of the work in the recent past was focused on the integration of SPC-based business methods and engineering analysis methods for performance assessment [5-8]. Unfortunately, the input data characteristics for these methods have not received much attention although the need for data characterization for use in metrics for benchmarking studies has been recognized [9].

The input data for the business analysis methods need not be high-frequency data that the engineering analysis methods use. Archived closed-loop data from the plant historian are typically used for these business analysis methods [10].

This work addresses the characterization and analysis of closed-loop archived process data used to calculate controller performance metrics in a statistical process control (SPC) framework with emphasis on manipulated variables in a multivariable control configuration (APC-MVs).

The organization of this paper is as follows. Section two introduces the concept of variability bands, which is an important characteristic of the actuating error time series. Section three describes four methods for error band identification using general purpose statistical tools like histograms, normal probability plots, quantile-quantile plots and the sample autocorrelation function; and the effect of sample time and data compression on these techniques. Section four discusses the effect of set-point variability on the error variability bands of manipulated variables in a multivariable controller configuration. Results are included in section three and four using data from a domestic petroleum refinery. Section five summarizes our findings and identifies areas of future work.

II. Error Variability Bands

Actuating error, defined as process variable minus the set-point, is a key variable for controller performance assessment. For single-input-single-output (SISO) loops, a well performing loop should reject disturbances and the process variable should closely follow the set-point. The variability in the controller error in such a case will be a direct indicator of controller performance [11]. For a multivariable control loop, however, the characteristics of the controller error can be different due to the presence of various factors including interactions between loops and constraints on certain variables.

Closed-loop archived data from a U.S. refinery are used in this paper to reveal the actuating error characteristics of manipulated variables employed by multivariable controllers. The controller configurations for each of the data sets are illustrated in the Appendix. Plots of the actuating error time series for the advanced process control manipulated variables (APC-MVs) exhibit different levels of variability when considered over a long period of time (one year). These intervals of variability are termed as "error variability bands." The variability bands are characterized by minimal within-band variation and considerable band-to-band variation.

Fig. 1 shows the actuating error time series from Oct 2003 to Sep 2004 for flow loop FC2, in which, at least two different bands can be identified.

The number of bands, total duration of the bands in days, and the standard deviation of the actuating error in these bands for flow, pressure and temperature examples of APC-MVs are tabulated in Table 1. Of the seven industrial closed-loop data sets involving APC-MVs, four of the sets had multiple error variability bands present and the other three had a single band. Anywhere between one to four bands have been typically observed in each of the error time series studied.

The error variability bands are due to the presence of assignable causes in the process. Examples include changing controller modes (APC on/off), feed stock variation, seasonal fluctuations and controller performance degradation. Assignable causes produce lost performance and therefore, if identified, provide an opportunity for improved performance.

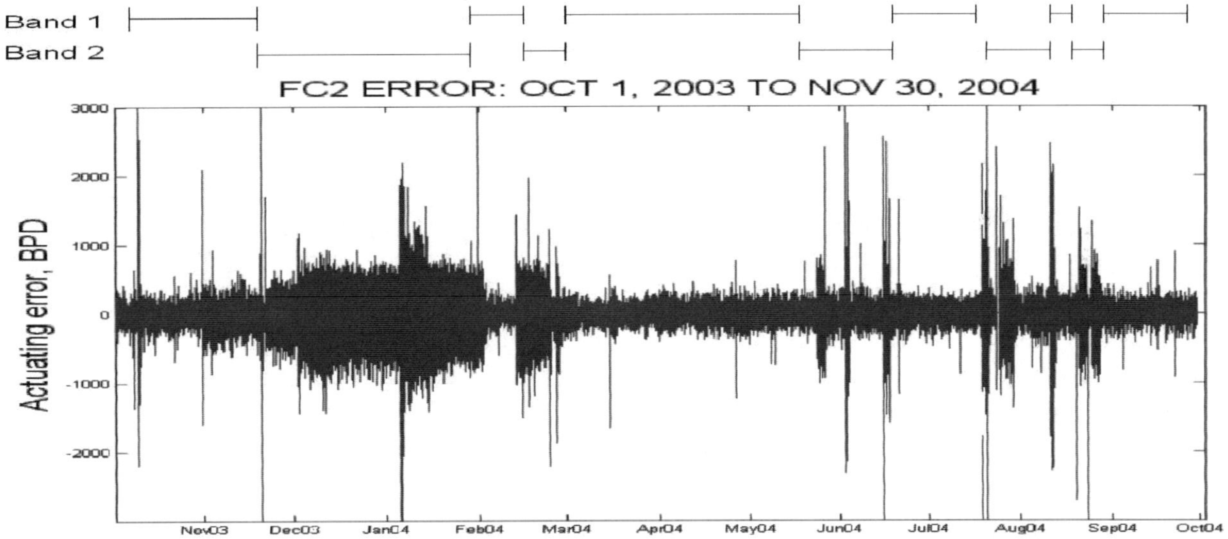

Figure 1: Actuating error time series for a flow loop indicating the presence of error variability bands.

Table 1: Error variability bands in the actuating error time series of the APC-MVs tabulated based on the visual observation of the time series.

Loop	#Bands	Band#	Days	STD	Units
FC1	1	1	360	35	bpd
FC3	3	1	328	89	bpd
		2	17	123	bpd
		3	15	126	bpd
PC1	2	1	273	0.04	psi
		2	39	0.06	psi
PC2	1	1	360	0.21	psi
PC4	1	1	360	0.18	psi
TC2	4	1	60	0.82	°F
		2	33	0.65	°F
		3	130	0.61	°F
		4	137	0.93	°F
TC3	4	1	60	0.48	°F
		2	30	0.37	°F
		3	171	0.43	°F
		4	99	0.45	°F

III. Error Variability Bands Identification

Automatic identification of the error variability bands provides the starting point for further diagnosis and elimination of assignable causes that can lead to real business improvement. This section presents four error variability band identification techniques using general purpose statistical tools including histograms, normal-probability plots, quantile-quantile plots and the sample autocorrelation function. The performance of these methods is presented using archived refinery data reconstructed on one-minute sample period for flow, pressure, and temperature loops.

Histogram

Histograms visually represent the frequency distribution of the data. A bell curve representing the normal probability density function, generated from the sample mean and sample standard deviation of the data, can be superimposed on the histogram to facilitate visual observation of deviation from normality.

The visual characteristics of histograms used in this paper are 'peak' and 'heavytails'. These terms represent deviation from the normal density curve at the center and at both the ends of the histogram, respectively. Fig. 2 shows the presence of a peak and heavytails in the histogram of an actuating error time series of a flow loop that has multiple error bands. When the intervals associated with the two error variability bands in Fig. 2 are plotted separately, the peak and the heavytails are removed as seen in Figs. 3 and 4. The peak and the heavytails observed in the Fig. 2 are a result of the overlap of distributions corresponding to the bands (indicated as intervals 1 and 2 in the Fig. 2).

Peak and heavytails were observed in the histograms of the actuating error time series that had error variability bands. Peak and heavytails were not found in the histograms of actuating error data that did not have error variability bands.

The disadvantage of the histogram technique is that it is best suited for only visual characterization, and does not lend itself to automation. The effect of outliers, and the choice of an inappropriate bin-size can effect the way the histogram appears, and hence the peak and heavytail characteristics.

The effect of compression on characteristics observed in the histogram is likely to be lessened because of the grouping of data into bins.

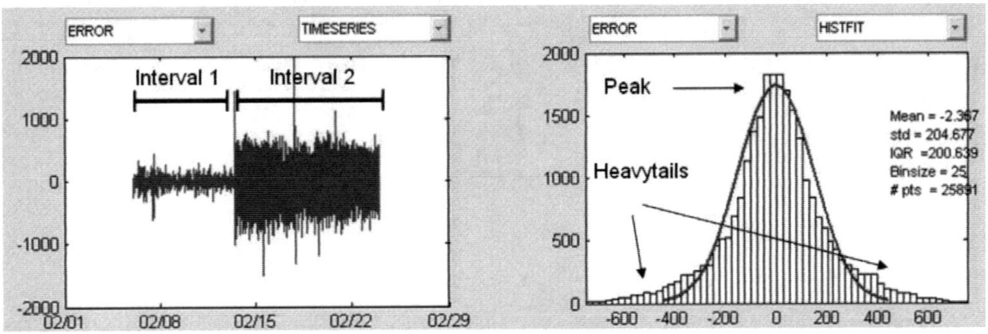

Figure 2: Illustration of the presence of 'peak' and 'heavytails' in the histogram of an actuating error series of FC2 flow loop for an interval that has two variability bands.

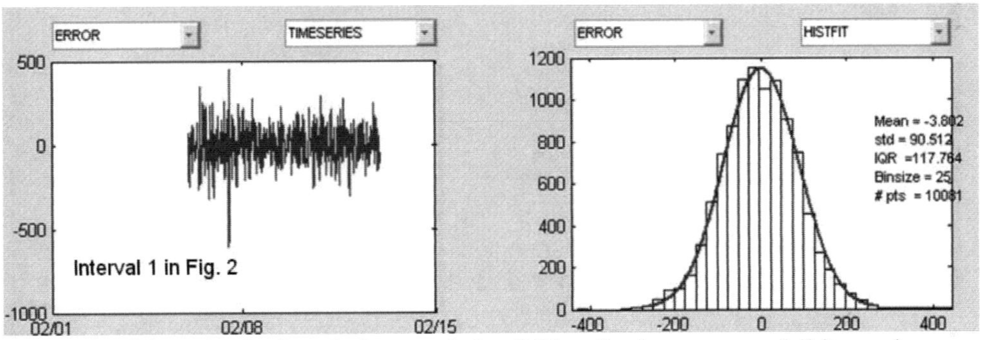

Figure 3: The single band, interval 1 of Fig. 2, does not exhibit peak or heavytails.

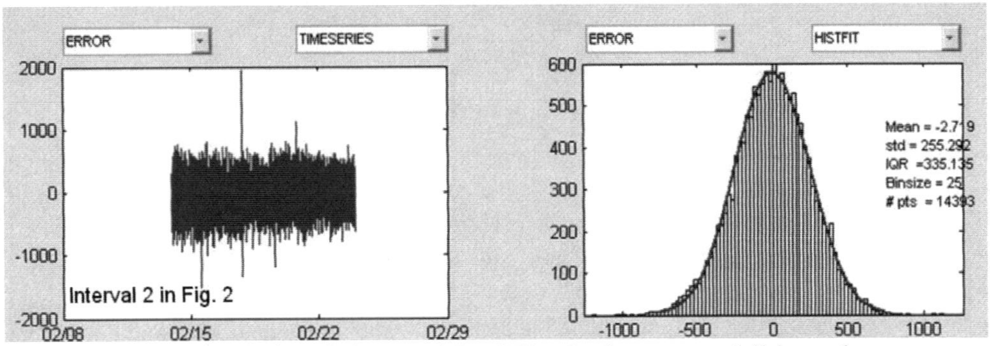

Figure 4: The single band, interval 2 of Fig. 2, does not exhibit peak or heavytails.

Normslope

Actuating error distributions showed close conformity to the normal distribution when there was a single band in the interval considered. A statistic based on the properties of a normal distribution is therefore proposed to characterize the band-spread.

The 'normslope' measures the spread of the data distribution based on the normal distribution property that the center 68% of the data will be contained in one standard deviation about either side of the mean. The normslope is calculated as the slope of the least-squares line through the center 68% of the data on a normal probability plot. The construction of a normal probability plot for a given data set can be found in [12]. Figure 5 illustrates the normal probability plot of a dataset of hundred points drawn randomly from a normal distribution of mean zero and standard deviation five. The slope of the linear fit through the center 68% of the data is obtained as the normslope as shown in Fig. 5.

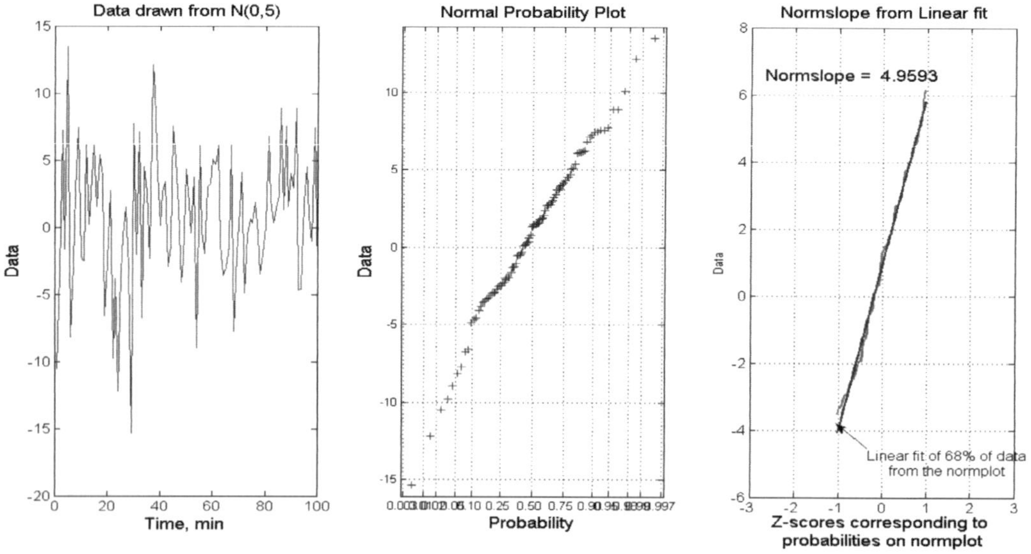

Figure 5: Normal probability plot and normslope of a dataset of 100 points drawn randomly from a normal distribution of mean zero and standard deviation five.

Table 2 shows the normslope for the APC-MVs for each month. It can be observed that for the loop FC3, normslope in the first two months is markedly different from the normslope calculated for other months of the year. In fact, four distinct bands can be identified in the table: Oct. and Nov. 2003, where the variability is about 130 bpd; Jan. through Apr. 2004, where the variability is about 80 bpd; Jun. through Aug. 2004, where the variability is about 95 bpd; and the months Dec 2003, May 2004 and Sep 2004 where the variability is about 103 bpd. These results are in agreement with visual observations of the time series as shown in Fig. 6. Error variability bands identified for the other loops in Table 2 also agree with the visual observations tabulated in Table 1.

The advantages of using the normslope include robustness to outliers since the center 68% of the data is considered, sensitivity to error variability bands, availability of spread estimates in engineering units, and simultaneous comparison of variability in different intervals. The disadvantage of using the normslope is that the spread estimate can be interpreted in engineering terms only when the deviations from normality are not significant. The effect of compression on the normslope statistic is going to be minimal as compression is unlikely to affect the amount of data that falls in the center 68% of the distribution.

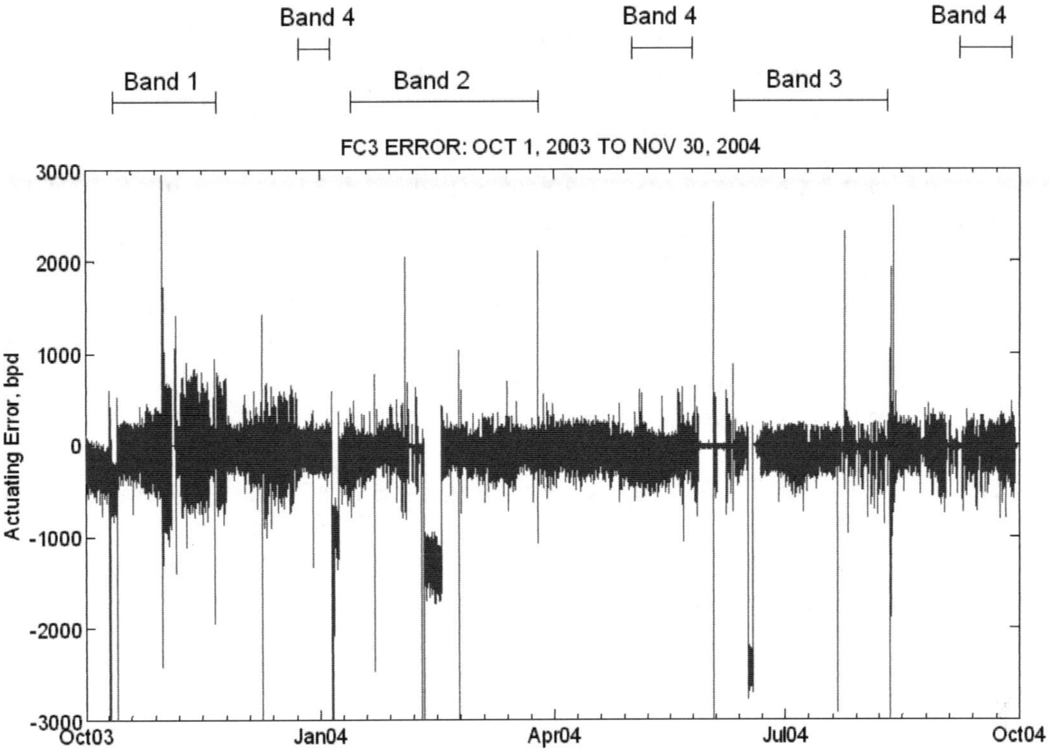

Figure 6: Actuating error time series for flow loop FC3. Figure verifies the presence of variability bands detected by the normslope statistic.

Table 2: Normslope statistic for the actuating error time series for various months for flow, pressure and temperature loops. This table, in conjunction with the actuating error time series, can be used to detect error variability bands.

Month	FC1	FC3	PC1	PC2	PC4	TC2	TC3
	bpd	bpd	psi	psi	psi	°F	°F
Oct03	35.63	138.74	0.05	0.27	0.18	0.88	0.42
Nov03	41.51	130.53	0.05	0.23	0.18	0.80	0.59
Dec03	38.87	103.22	0.06	0.22	0.17	0.66	0.40
Jan04	34.09	80.56	0.07	0.19	0.18	0.64	0.45
Feb04	34.42	81.56	0.05	0.19	0.18	0.68	0.50
Mar04	-	82.93	-	0.21	0.18	0.59	0.29
Apr04	34.68	83.01	0.04	0.19	0.16	0.63	0.22
May04	34.49	103.74	0.04	0.18	0.19	0.78	0.26
Jun04	33.55	95.47	0.04	0.22	0.18	1.00	0.50
Jul04	33.30	97.27	0.04	0.24	0.19	0.92	0.26
Aug04	32.35	95.90	0.05	0.21	0.19	1.03	0.42
Sep04	35.33	101.30	0.05	0.20	0.17	1.19	0.49

qqslope

The statistic 'qqslope', analogous to normslope, is ratio of the slope of the least-squares line through the first and third quartiles of the sampled distribution and the slope of the least-squares line through the first and third quartiles of the annual (or reference) distribution when plotted on a quantile-quantile plot. The construction of the quantile-quantile plot is described in [13] and the calculation of the qqslope is similar to the normslope calculation described in the previous section.

The qqslope will be an indication of whether the variability of the sampled distribution is greater or smaller than the annual (or reference) distribution. For example, if the qqslope is greater than 1.0, then the sampled distribution has greater spread than annual distribution.

The qqslope ratio thus allows for comparing the spread of the monthly distributions to the annual distribution. If the annual distribution is considered as the 'average' characteristic of the loop, then the monthly distribution will determine the deviation from the annual average.

As an example, consider loop FC3. As indicated in the Table 2, at least four bands can be identified in FC3: Oct. 2003 and Nov. 2003, where the variability in actuating error is about 1.3 to 1.4 times the annual; Jan. 2004 through Apr. 2004, where the variability in the actuating error is about 0.85 times the annual; Jun. 2004 through Aug. 2004 where the variability in the actuating error is about the same as the annual; and in the months Dec 2003, May 2004 and Sep 2004 where the variability is 1.06 times annual. This result is consistent with visual observation of the FC3 error time series and the normslope results presented in the previous section.

The qqslope can be used to compare two distributions without any reliance on a standard model. The qqslope characterization can be used, for instance, to compare the error distribution for each month to the composite distribution for a full year. The idea is to evaluate the current and long-term controller performance. In this case, the annual composite distribution can be considered as the reference or benchmark distribution.

Similar to normslope, the effect of compression on the qqslope statistic is going to be minimal as compression is unlikely to affect the amount of data that falls in each quantile of the data.

Auto Correlation Function (ACF)

The peak and heavytails, normslope and the qqslope characteristics described so far involve sorting or grouping of data with total disregard to the order of the time series. These techniques also consider the data to be uncorrelated. A statistic based on the shape of the autocorrelation function is proposed as a complementary method to the above techniques as it provides a fundamentally different approach based on the order of the time series.

The (non-zero) ACF coefficients are a measure of the randomness of the time series [14]. For a data sample, the sample ACF coefficients are never zero, but lie within a confidence interval about zero. Initial results indicate that the number of appreciable autocorrelation coefficients can be used to detect the error variability bands. The results and the effectiveness of the ACF technique will be presented in a future paper.

The ACF method takes all the data into account and is likely to be most impacted by the presence of outliers.

Table 3: qqslope statistic for the actuating error time series for various months for flow, pressure and temperature loops. The reference distribution for here is the annual distribution whose qqslope is 1.0

Month	FC1	FC3	PC1	PC2	PC4	TC2	TC3
Oct03	0.97	1.42	0.89	1.29	1.04	1.11	1.12
Nov03	1.13	1.34	0.89	1.07	0.99	1.01	1.57
Dec03	1.06	1.06	1.13	1.04	0.94	0.83	1.05
Jan04	0.93	0.83	1.32	0.89	0.99	0.81	1.19
Feb04	0.94	0.84	0.88	0.90	1.01	0.86	1.34
Mar04	-	0.85	-	0.99	1.03	0.75	0.77
Apr04	1.02	0.85	0.96	0.91	0.92	0.79	0.58
May04	1.02	1.06	0.97	0.85	1.04	0.99	0.69
Jun04	0.99	0.98	1.01	1.04	0.99	1.26	1.35
Jul04	0.98	1.00	0.92	1.14	1.05	1.17	0.69
Aug04	0.95	0.98	1.22	0.99	1.04	1.30	1.13
Sep04	1.04	1.04	1.23	0.96	0.96	1.50	1.32

Band 1 Band 2 Band 3 Band 4

IV. Impact of set-point variability on APC-MVs

One of the goals of a multivariable controller is to transfer variability from the controlled variables (CVs) to the manipulated variables (MVs), within the process constraints. To achieve this, the multivariable controller constantly adjusts the set-points of the MVs to keep CV variability to a minimum. Additional set-point variability, characterized by frequent set-point changes, is introduced in the APC-MVs when the multivariable controller is turned on.

This section considers the impact of this additional set-point variability on the actuating error time series of the APC-MVs in an advanced control configuration.

Figures 7-9 show the actuating error time series for a flow, pressure and temperature loop during intervals of different set-point variability. When the sample standard deviation was used as the measure of variability, the following results were obtained: For TC3 (Fig. 7), the set-point variability increased from zero to 1.4 °F when the multivariable controller was turned on, and the error variability increased slightly from 0.52 °F to 0.58 °F. For PC2 (Fig. 8), the set-point variability increased from 0.2 psi to 0.9 psi and the error variability doubled from 0.16 psi to 0.32 psi. For FC3 (Fig. 9), the set-point variability decreased from 115.8 bpd to zero, and the error variability decreased from 92 bpd to 41 bpd. In each of the cases, the actuating error variability increased or decreased as the set-point variability increased or decreased.

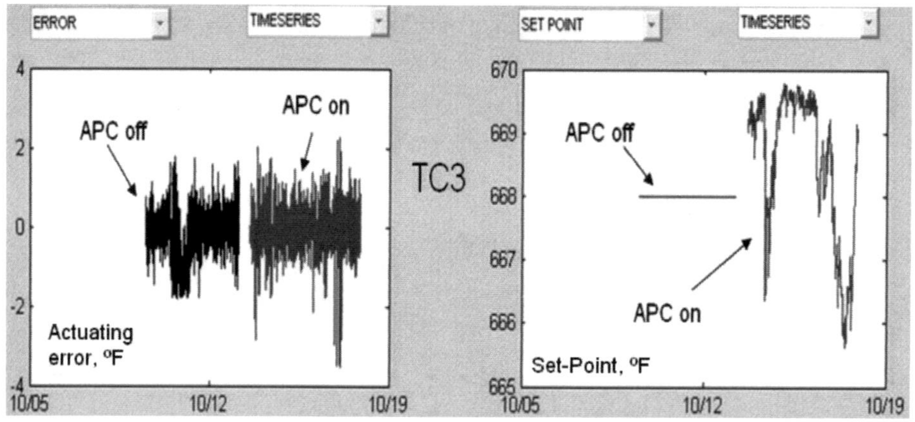

Figure 7: Error variability increase in a (APC-MV) temperature loop TC3 owing to an increase in the set-point variability when APC is turned on.

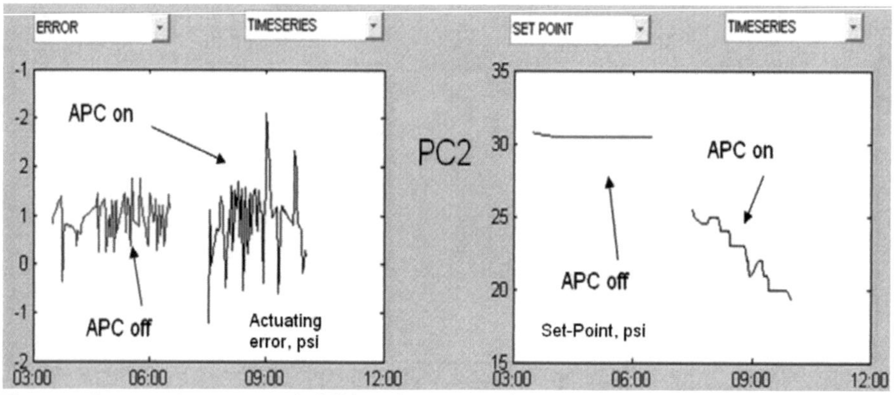

Figure 8: Error variability increase in a (APC-MV) pressure loop PC2 owing to an increase in the set-point variability when APC is turned on.

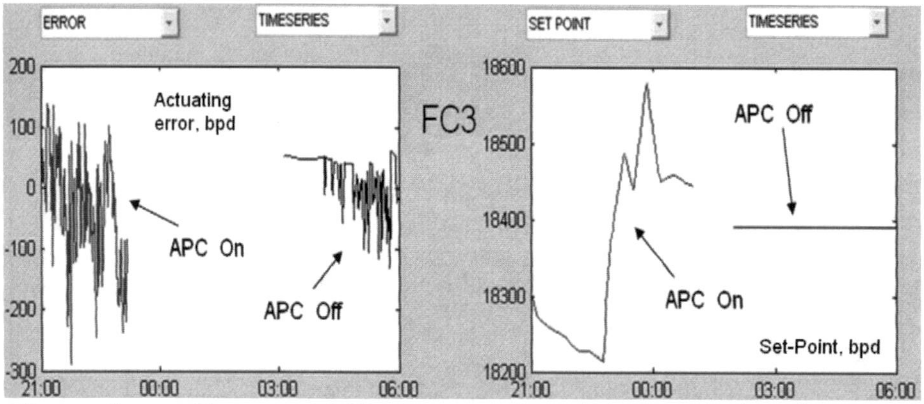

Figure 9: Error variability increase in a (APC-MV) flow loop FC3 owing to an increase in the set-point variability when APC is turned on.

Contributions to the actuating error variability:

The actuating error distribution is the resultant joint-distribution of the process variable and the set-point distributions, described by equation (1). The actuating error variability, therefore, depends on the set-point variability (APC action) and on the SP-PV co-variability (closed-loop dynamics). That is:

$$error = PV - SP \tag{1}$$

$$\gamma_{error} = \gamma_{PV} + \gamma_{SP} - 2\gamma_{SP,PV} \tag{2}$$

Where [15]:

γ_{error} = variance of the error, γ_{PV} = variance of PV, γ_{SP} = variance of SP and

$\gamma_{SP,PV}$ = covariance of SP and PV.

Re-writing equation (2) in terms of correlation,

$$\rho_{SP,PV} = \frac{\gamma_{SP,PV}}{\sqrt{\gamma_{PV}\gamma_{SP}}} \tag{3}$$

we obtain:

$$\gamma_{error} = \gamma_{PV} + \gamma_{SP} - 2\rho_{SP,PV}\sqrt{\gamma_{PV}\gamma_{SP}} \tag{4}$$

Standard deviation (square root of variance) and correlation estimates of the set-point, the process variable and actuating error are defined as indicated below in equations (5) – (9):

$$\hat{\sigma}_{SP} = \sqrt{\frac{\sum_{i=1}^{N}(SP_i - \overline{SP})^2}{N-1}} \tag{5}$$

$$\hat{\sigma}_{PV} = \sqrt{\frac{\sum_{i=1}^{N}(PV_i - \overline{PV})^2}{N-1}} \tag{6}$$

$$\hat{\sigma}_{SP,PV} = \sqrt{\frac{\sum_{i=1}^{N}(SP_i - \overline{SP})(PV_i - \overline{PV})}{N-1}} \tag{7}$$

$$\hat{\sigma}_{error} = \sqrt{\frac{\sum_{i=1}^{N}(error_i - \overline{error})^2}{N-1}} \tag{8}$$

$$\hat{\rho}_{SP,PV} = \frac{\hat{\sigma}^2_{SP,PV}}{\hat{\sigma}_{SP}\hat{\sigma}_{PV}} \tag{9}$$

Substituting the estimates into equation (4) yields the back-calculated error variance:

$$\tilde{\sigma}^2_{error} = \hat{\sigma}^2_{SP} + \hat{\sigma}^2_{PV} - 2\hat{\sigma}_{SP}\hat{\sigma}_{PV}\rho_{SP,PV} \tag{10}$$

The PV-SP correlation $\hat{\rho}_{SP,PV}$ in equation (10) captures the closed-loop dynamics of the process. The correlation, $\hat{\rho}_{SP,PV}$, is a measure of the controller performance with $\hat{\rho}_{SP,PV} = 1.0$ representing perfect control, and $\hat{\rho}_{SP,PV} = 0.0$ representing no control.

Perfect control: PV=SP implies $\hat{\sigma}_{PV} = \hat{\sigma}_{SP}$, and $\hat{\rho}_{SP,PV} = 1$. Substitution in equation (10) results in a zero error variability as the individual variability components cancel.

No control: PV and SP are independent, and $\hat{\rho}_{SP,PV} = 0$. Equation (10) simplifies to:

$$\tilde{\sigma}^2_{error} = \hat{\sigma}^2_{SP} + \hat{\sigma}^2_{PV} \tag{11}$$

The error variance in this case is the sum of the variances of the SP and PV, respectively.

If the frequency of set-point changes is faster than the rate at which the process can respond, the controller performance degrades and results in higher actuating error variability. In other words, this means that high frequency of the set-point changes leads to an increased SP variability and decreased $\hat{\rho}_{SP,PV}$ resulting in an increase in the actuating error variability according to equation (10).

We are analyzing the data to identify the threshold beyond which an increase in the frequency of set-point changes negatively impacts the actuating error variability.

Effect of variability bands on variance estimators:

The following discussion examines the application of equation (10) to the data and the limitations of the types of estimators given in equations (5) – (9) from which equation (10) was derived.

The estimates of the standard deviations for two intervals each of FC1 and TC2 are tabulated in Table 4 using equations (5) – (9). For example the standard deviation estimates for FC1 from 25-Oct-2003 through 28-Oct-2003, from Table 4 are:

$$\hat{\sigma}_{PV} = 186 \text{ bpd}, \ \hat{\sigma}_{SP} = 182 \text{ bpd}, \ \hat{\rho}_{SP,PV} = 0.98, \text{ and } \hat{\sigma}_{error} = 36 \text{ bpd}.$$

The high SP-PV correlation is an indication of good control in this interval. The back-calculated actuating error variability from individual variability components obtained from equation (10) as follows:

$$\tilde{\sigma}^2_{error} = 182^2 + 186^2 - 2(182)(186)(0.98) \Rightarrow \tilde{\sigma}_{error} \approx 36 \text{ bpd}$$

The standard deviation estimates of the actuating error obtained by back-calculation from individual variability components and directly from the actuating error data for all the intervals listed in Table 4 are tabulated in Table 5. It can be observed from Table 5 that the predictions from equation (10) are in good agreement for the interval: FC1 from 25-Oct-2003 to 28-Oct-2003 and in poorer agreement for the two intervals: FC1 from 16-Apr-2004 to 5-May-2004 and TC2 from 9-Jun-2004 to 15-Jun-2004.

Figures 10-12 show the actuating error time series and the set-point time series for the intervals considered in Tables 4 and 5. For FC1 from 25-Oct-2003 to 28-Oct-2003 (Fig. 10), the multivariable controller is on, and the set-point is constantly varying but the actuating error time series does not contain multiple bands or outliers.

For FC1 from 16-Apr-2004 to 5-May-2004 (Fig. 11), the multivariable controller is on, and the set-point is constantly varying. The actuating error time series does not contain multiple variability bands but two unusually high values greater than 500 bpd are present in the error series.

For TC2 (Fig. 12) from 9-Jun-2004 to 15-Jun-2004, the multivariable controller is on, and the set-point is constantly varying. The actuating error time series exhibits multiple variability bands in this interval.

Based on the results from Table 5 and Figures 10-13, it is seen that equation (10) does not predict accurately in the presence of multiple variability bands and outliers. This is an important result since SPC-based performance analysis assumes a longer-term view over which the data are likely to contain multiple variability bands or outliers.

Table 4: Estimates for the standard deviations and correlation for various flow, pressure and temperature loops.

Loop	Start	End	APC STATUS	$\hat{\sigma}_{SP}$	$\hat{\sigma}_{PV}$	$\hat{\rho}_{SP,PV}$	$\hat{\sigma}_{error}$	Units
FC1	25-Oct-03	28-Oct-03	APC ON	182.0	186.0	0.98	36.0	bpd
FC1	16-Apr-04	5-May-04	APC ON	268.0	272.0	0.98	56.0	bpd
TC2	9-Jun-04	15-Jun-04	APC ON	2.6	2.9	0.89	1.3	°F

Table 5: Values of error variability (standard deviation) for back-calculated error variability from individual variability components $(\tilde{\sigma}_{error})$ and the error variability estimated directly from error data $(\hat{\sigma}_{error})$.

Loop	Start	End	Error back-calculated from individual variability components $\tilde{\sigma}_{error}$	Estimated from error data directly $\hat{\sigma}_{error}$	Relative error, %	Units
FC1	25-Oct-03	28-Oct-03	36.0	36.0	0	bpd
FC1	16-Apr-04	5-May-04	37.0	56.0	51	bpd
TC2	9-Jun-04	15-Jun-04	1.7	1.3	24	°F

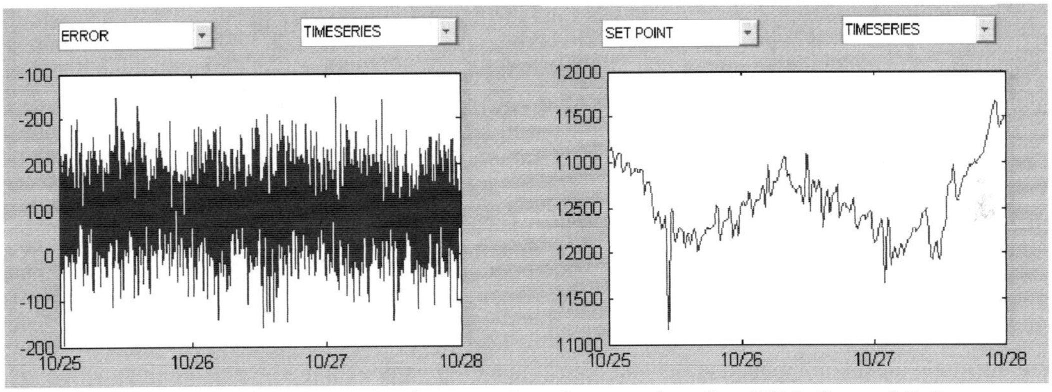

Figure 10: Actuating error and set-point time series for FC1 from 25-Oct-2003 to 28-Oct-2003. The multivariable controller is on during this period.

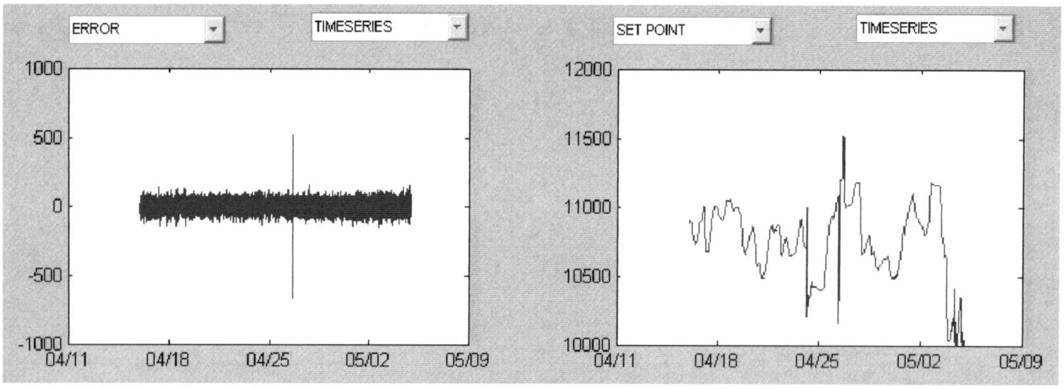

Figure 11: Actuating error and set-point time series for FC1 from 16-Apr-2004 to 5-May-2004. The multivariable controller is on during this period.

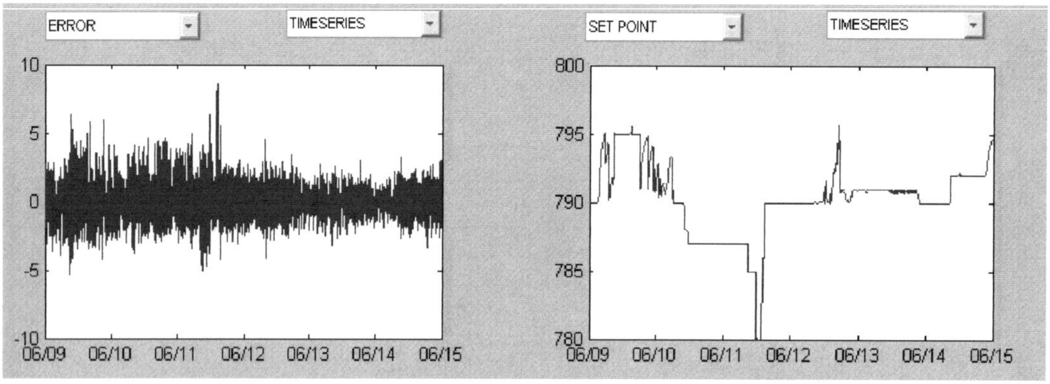

Figure 12: Actuating error and set-point time series for TC2 from 9-Jun-2004 to 15-Jun-2004. The multivariable controller is on during this period.

Implications on SPC-based business analysis:

The results presented in the preceding section can be useful in the interpretation of SPC-metrics used for business analysis purposes. The presence of peak and heavytails in the actuating error histograms is an indication of the presence of multiple variability bands in the interval considered. These multiple variability bands affect the SPC-based metrics and their interpretation. The multiple variability bands reveal performance changes in the controller and thereby provide opportunities for business improvement.

The idea of comparing error distributions in two periods is not sufficient, especially when dealing with multivariable controller configurations. The set-point variability and the mode in which the controller operates (APC on/off) also influence the actuating error variability.

V. Conclusions and Future Work

For the seven APC-MV datasets considered, normslope and qqslope were found to be robust and effective techniques for the quantification of error variability bands. Error variability bands detected using the normslope and the qqslope techniques agree well with visual observations for all the datasets.

Additional variability introduced in the actuating error series of APC-MVs due to an increase in the set-point variability (characterized by an increase in the frequency of set-point changes) depends on two factors – the change in set-point variability, and the closed-loop dynamics.

Although this paper is focused on APC manipulated variables, the results of this work are also applicable to the analysis of traditional, non-APC control systems. Initial results indicate that the closed-loop archived data characterization can improve our understanding of business analysis metrics.

Joint distributions models for actuating error using theoretical and empirical process variable and set-point distributions would lead to a further understanding of the transfer of set-point variability to the actuating error variability. The results from this work provide a platform for the development of such models.

References

[1] R. Christian, "Towards a common benchmark for long-term process control and monitoring performance evaluation," *Water Science and Technology*, vol. 50, pp. 41-49, 2004.

[2] J. T. Herman, "Capability Index - Enough for process industries?" presented at ASQC Quality Congress Transactions, Toronto, 1989.

[3] P. Lant and M. Steffens, "Benchmarking for process control: "Should I Invest in Improved Process Control?"" *Water Science and Technology*, vol. 37, pp. 49-54, 1998.

[4] G. Box and T. Kramer, "Statistical Process Monitoring and Feedback Adjustment: A Discussion," *Technometrics*, vol. 34, pp. 251-267, 1992.

[5] T. Kourti and J. F. MacGregor, "Multivariate Statistical Process Control and Industrial Applications," presented at Quality Congress. ASQ's Annual Quality Congress Proceedings, 2004.

[6] Y. Ling and S.-H. Sheu, "Integrating EPC and SPC for MIMO system," presented at IEEE International Conference on Systems, Man and Cybernetics, New York, 2005.

[7] J. F. MacGregor, "Using On-line Process Data to Improve Quality: Challenges for Statisticians," *International Statistical Review*, vol. 65, pp. 309-323, 1997.

[8] W. T. Tucker, F. W. Faltin, and V. A. Wiel, "Algorithmic Statistical Process Control: An Elaboration," *Technometrics*, vol. 35, pp. 363-375, 1993.

[9] A. Jansky, "Learning from the best with process benchmarking in nuclear power plants," presented at International Conference on Nuclear Engineering, Miami, Florida, 2006.

[10] R. D. De Veaux, L. H. Ungar, and J. M. Vinson, "Statistical Approaches to Fault Analysis in Multivariate Process Control," presented at Proceedings of the American Control Conference, Baltimore, Maryland, 1994.

[11] N. F. Thornhill, M. Oettinger, and P. Fedenczuk, "Refinery-wide control loop performance assessment," *Journal of Process Control*, vol. 9, pp. 109-124, 1999.

[12] N. A. Weiss, *Introductory Statistics*, 4th edition ed: Addison-Wesley Publishing Company, Inc., 1995.

[13] R. K. Pearson, "Exploring Process Data," *Journal of Process Control*, vol. 11, pp. 179-194, 2001.

[14] C. Chatfield, *The analysis of time series, an introduction*, 4th ed. NY: Chapman and Hall, 1989.

[15] O. J. Dunn and V. A. Clark, *Applied Statistics: Analysis of Variance and Regression*, 2nd edition ed: John Wiley & Sons, 1987.

Appendix

This appendix illustrates the configuration of the APC-MVs that were used for characterization studies.

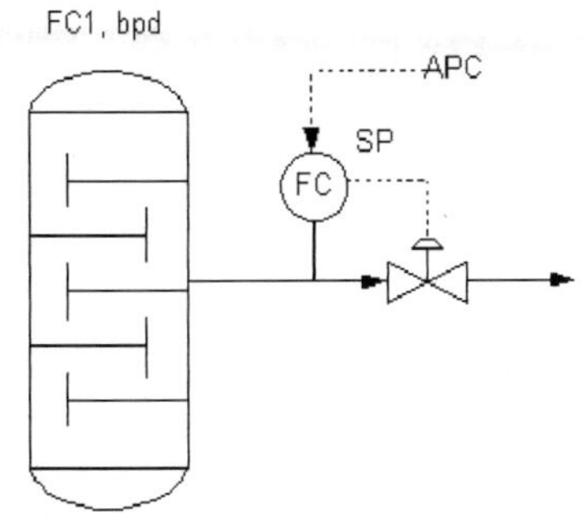

FC1, bpd. PV is an APC MV. The output from the APC contrilloer is the set point to the FC loop.

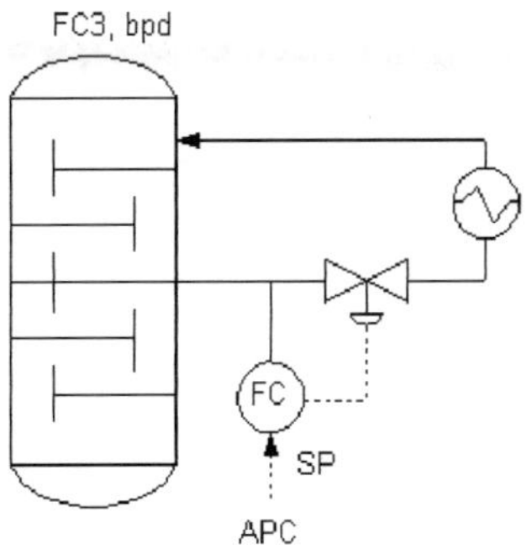

FC3, bpd. PV is an APC MV. The output from the APC controller is the set point to the FC loop.

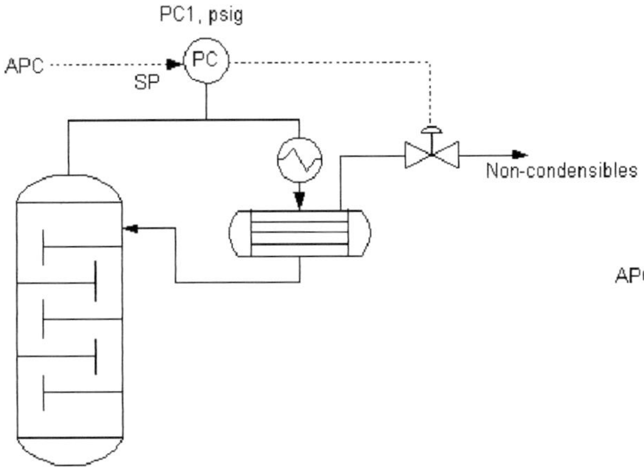

PC1, psig. PV is an APC MV. The output from the APC controller is the set point to the PC.

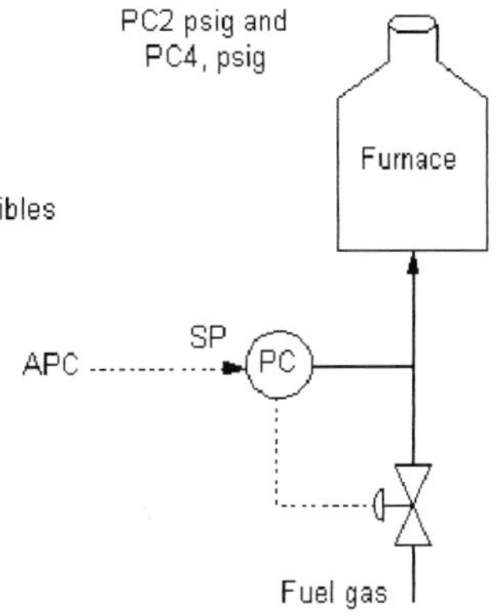

PC2 psig and PC4, psig. PV is an APC MV. The output from the APC controller is the set point to the PC.

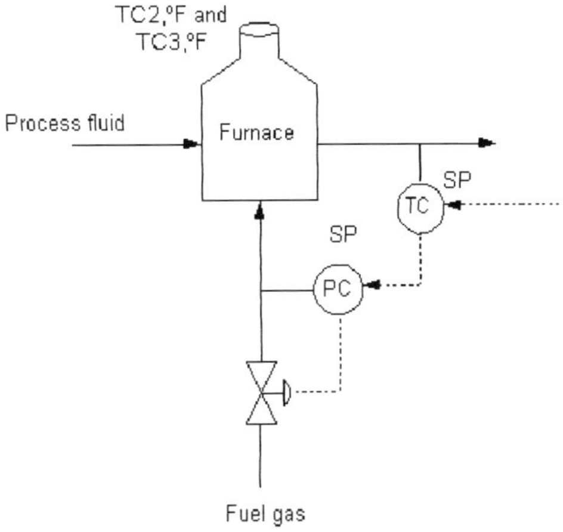

TC2°F, and TC3, °F. PV is an APC MV. The output from the APC controller is the set point to the TC.

Process Monitoring and Parameter Estimation via Unscented Kalman Filtering

Cheryl C.Y. Qu and Juergen Hahn. Department of Chemical Engineering, Texas A & M University, College Station, TX 77843-3122

1 Introduction

Extended Kalman filters (EKF) have found wide-spread use in nonlinear state and parameter estimation. Unlike EKF, where the first-order linearization is used for error covariance estimation, unscented Kalman filters (UKF), as proposed by Julier and Uhlman [1], pass the mean and covariances of estimated states through a nonlinear transformation. This is achieved by carefully choosing a set of sigma points, which capture the true mean and covariance of the given distribution and results in UKF being capable of estimating the posterior mean and covariances accurately to a high order. However, in theory, performance of UKF can exceed the one for EKF where only first-order accuracy is achieved. Despite UKF's potential for improved state and parameter estimation, only few applications in chemical engineering have been reported [2][3][4].

In this work, UKF has been applied to a CSTR [5], in order to investigate the performance achieved by the state and parameter estimator and compared to results generated by EKF.

The case study shows that UKF performs as well as EKF when the CSTR is operated in a locally linear operating region. However, when the magnitude of the operating region is enlarged, UKF performs slightly better than EKF since the nonlinear behavior becomes more important. It is also exemplified that UKF can offer benefits over EKF if the operating conditions of a process vary drastically. Also, the parameter estimates achieved by UKF tend to be more accurate than the one computed by EKF for more extreme operating conditions.

This paper is organized as follows. In Section 2, a brief review of the problem of nonlinear state and parameter estimation is presented along with one of the most widely used EKF algorithm. Section 3 presents the UKF algorithms for the nonlinear estimation problems in chemical engineering, followed by Section 4 which shows the performance of each filter for state and parameter estimation. Concluding remarks are given in Section 5.

2 Preliminaries

2.1 State Estimation

A class of nonlinear discrete time system with measurements is given by:

$$\begin{aligned} x_k &= f(x_{k-1}, u_{k-1}, w_{k-1}) \\ y_k &= h(x_k, u_k, v_k) \end{aligned} \tag{1}$$

where $x_k \in \mathbb{R}^n$ is a vector of state variables, the function f and h are discrete-time differentiable functions of the state vector x and input u respectively. $w_k \in \mathbb{R}^n$ is a vector of plant noise, with $E[w_k] = 0$ and $E[w_k w_k^T] = Q_k$; $y_k \in \mathbb{R}^m$ is a vector of measured variables, the function h is a continuously differentiable function of the state vector x and $v_k \in \mathbb{R}^m$ is a vector of measurement noise, with $E[v_k] = 0$ and $E[v_k v_k^T] = R_k$; n is the number of states, m refers to the number of measurement variables. The distributions of w and v are not necessary Gaussian. The initial value x_0 may be assumed to be a Gaussian random variable with known mean and known $n \times n$ covariance matrix P_0.

The objective is to find an estimate $\hat{x}_k$ of x_k to minimize the weighted mean-squared error $E(x_k - \hat{x}_k)M(x_k - \hat{x}_k)^T$, where M is any symmetric nonnegative definite weighting matrix. If all estimates weight equally, the objective becomes to minimize the error covariance matrix for an unbiased estimator, given by $P = E(x_k - \hat{x}_k)(x_k - \hat{x}_k)^T$. More specifically, the trace of P is chosen to be minimized resulting in the performance index $J = \frac{1}{2}\mathbf{Tr}[E(x_k - \hat{x}_k)(x_k - \hat{x}_k)^T]$.

2.2 Parameter Estimation

The parameter estimation problem involves determining a nonlinear mapping which is of the form:

$$\begin{aligned} x_k &= f(x_{k-1}, u_{k-1}, w_{k-1}, p) \\ y_k &= h(x_k, u_k, v_k) \end{aligned} \tag{2}$$

where p is a scalar or vector parameterizing the nonlinear function f. The description of p corresponds to a stationary process with identity state transition matrix, driven by process noise w_{k-1}.

One way of estimating parameters is to augment the state vector with parameters to be estimated: $z_k = [x_k^T \ p_k^T]^T$. The estimation of both states and parameters can be done recursively by writing the state-space representation as :

$$z_k = \begin{pmatrix} f(x_{k-1}, u_{k-1}, w_{k-1}, p_{k-1}) \\ p_{k-1} + w_{k-1} \end{pmatrix}. \tag{3}$$

2.3 Extended Kalman Filter

The equations defining the discrete-time form of the extended Kalman filter are summarized as follows:

Prediction equations:

$$\begin{aligned} \hat{x}_{k|k-1} &= f(\hat{x}_{k-1|k-1}, u_{k-1}) \\ \hat{y}_k &= h(\hat{x}_{k|k-1}, u_k) \end{aligned} \tag{4}$$

Update equations:

$$\begin{aligned} P_{k|k-1} &= A_{k-1} P_{k-1|k-1} A_{k-1}^T + G_{k-1} Q G_{k-1}^T \\ K_k &= P_{k|k-1} C_k^T (C_k P_{k|k-1} C_k^T + H_k R H_k^T)^{-1} \\ P_{k|k} &= (I - K_k C_k) P_{k|k-1} \\ \hat{x}_{k|k} &= \hat{x}_{k|k-1} + K_k (y_k - \hat{y}_k) \end{aligned} \tag{5}$$

where $A_{k-1} \approx \frac{\partial f}{\partial x}|_{\hat{x}_{k-1}^-}$, $C_k \approx \frac{\partial h}{\partial x}|_{\hat{x}_{k|k-1}}$, $G_{k-1} \approx \frac{\partial f}{\partial w}|_{w_{k-1}}$ and $H_k \approx \frac{\partial h}{\partial v}|_{v_k}$ are the matrices of the linearized system model and computed in real time as functions of the estimate for linearization about the estimated trajectory.

3 Process Monitoring via Unscented Kalman Filter

A Kalman filter is the optimal solution for the linear estimation problems. In order to apply a Kalman filter to nonlinear estimation problems, a linearization method can be used to get approximate solutions [6]. Linearized Kalman filter (LKF) has low computation cost since the system state transition matrix can be computed off-line. However, the deviation of the actual values from the initial states or steady states tends to be large when input changes or state perturbations occur. Schmidt proposed the Extended Kalman filter (EKF) by linearizing about the estimated values at each instant [7]. EKF has become the most widely used estimator for nonlinear systems. However, EKF, just as LKF, still does not avoid two well-known drawbacks: Firstly, highly unstable filters can be produced by linearization if the assumptions of local linearity is violated. Secondly, the Jacobian matrix does not always exist since some systems contain discontinuities or deriving Jacobian matrix could be computationally expensive. To overcome the limitation by linearization, Julier *et. al* [1] proposed the Unscented transformation (UT), by which both mean and covariance are propagated through nonlinear transformations.

Unscented Kalman filter is the application of the unscented transformation (UT) to recursive estimation. The main point is to use UT to address the deficiencies of linearization by providing a more direct and explicit mechanism for transforming mean and covariance information. The prediction algorithm using the unscented transform is given in [8] but may need extensions and modifications for a specific application. [3] and [4] have presented the UKF algorithm and studied the UKF applications in chemical engineering. However, it needs to be pointed out that only those sigma points contributing to the system states will be mapped through the nonlinear systems. Therefore, the explicit UKF algorithm for a dynamic system subject to process and observation noise is presented in this section.

In the UT mechanism, a set of weighed sigma points are deterministically chosen from the statistics of the transformation so that certain properties of these points (e.g. a given mean and covariance) match those of the prior distribution. These sigma points are propagated through a nonlinear mapping and then a weighted mean and covariance is computed. One way to find one set of sigma points that are accurate to the second order is described as follows:

1. Augmenting the system state vector to an $n^a = n+q+r$ dimensional vector $x^a = [x^T\ w^T\ v^T]^T$, its mean and covariance are

$$\begin{aligned} \hat{x}^a_{k-1|k-1} &= \begin{pmatrix} \hat{x}_{k-1|k-1} \\ \mathbf{0^{q\times 1}} \\ \mathbf{0^{r\times 1}} \end{pmatrix} \\ P^a_{k-1|k-1} &= \begin{pmatrix} P_{k-1|k-1} & \mathbf{0^{n\times q}} & \mathbf{0^{n\times r}} \\ \mathbf{0^{q\times n}} & Q_{k-1} & P^{wv}_{k-1} \\ \mathbf{0^{r\times n}} & P^{vw}_{k-1} & R_{k-1} \end{pmatrix} \end{aligned} \tag{6}$$

where n is the dimension of system original state vector, q and r are the dimensions of the system and measurement noise vectors and P^{vw}_{k-1} and P^{wv}_{k-1} are the correlations between the system and measurement noises. For ease of computation, P^{vw}_{k-1} and P^{wv}_{k-1} are usually set to zero.

2. Generating a set of $2n^a+1$ symmetric sigma points

$$\chi^a_{k-1} = \hat{X}^a_{k-1|k-1} + \begin{pmatrix} \mathbf{0} & \sqrt{(n^a+\kappa)P^a_{k-1|k-1}} & -\sqrt{(n^a+\kappa)P^a_{k-1|k-1}} \end{pmatrix} \tag{7}$$

where $\hat{X}^a_{k-1|k-1}$ is the expanded $n^a \times (2n^a+1)$ matrix with $\hat{x}^a_{k-1|k-1}$ as each column. $\kappa \in \Re$ is a parameter providing flexibility to fine tune the higher order moments of the approximation. The more higher order moments are taken into account, the less the overall prediction error will be [8]. For a Gaussian distribution $x(k)$, a useful heuristic is to select $n^a+\kappa = 3$.

The essential equations defining the unscented Kalman filter are summarized as follows:

Prediction equations:

$$
\begin{aligned}
\chi_k^x &= f(\chi_{k-1}^x, \chi_{k-1}^w, u_{k-1}) \\
\hat{x}_{k|k-1} &= \sum_{i=1}^{2n^a+1} W_i \chi_{i,k}^x \\
\gamma_k &= h(\chi_k^x, \chi_{k-1}^v, u_k) \\
\hat{y}_k &= \sum_{i=1}^{2n^a+1} W_i \gamma_{i,k}
\end{aligned} \tag{8}
$$

Update equations:

$$P_{k|k-1} = \sum_{i=1}^{2n^a+1} W_i [\chi_{i,k}^x - \hat{x}_{k|k-1}][\chi_{i,k}^x - \hat{x}_{k|k-1}]^T \tag{9}$$

$$P_{y,k} = \sum_{i=1}^{2n^a+1} W_i [\gamma_{i,k} - \hat{y}_k][\gamma_{i,k} - \hat{y}_k]^T \tag{10}$$

$$P_{xy,k} = \sum_{i=1}^{2n^a+1} W_i [\chi_{i,k}^x - \hat{x}_{k|k-1}][\gamma_{i,k} - \hat{y}_k]^T \tag{11}$$

$$K_k = P_{xy,k} P_{y,k}^{-1} \tag{12}$$

$$P_{k|k} = P_{k|k-1} - K_k P_{y,k} K_k^T \tag{13}$$

$$\hat{x}_{k|k} = \hat{x}_{k|k-1} + K_k (y_k - \hat{y}_k) \tag{14}$$

where $\chi^a = [(\chi^x)_{1 \times n}^T \ (\chi^w)_{1 \times q}^T \ (\chi^v)_{1 \times r}^T]^T$ and W_i are weights as calculated in Eq. (15)

$$W_i = \begin{cases} \frac{\kappa}{2(n_a+\kappa)}, & \text{if i=1;} \\ \frac{1}{2(n_a+\kappa)}, & \text{otherwise.} \end{cases} \tag{15}$$

It is worth noting that if process noise or observation noise enter the system and measurements linearly, the system states and its covariance matrix need not to be augmented as in Eq. (6). However, Q and R need to be added linearly to Eq. (9) and (10) respectively.

The properties that a set of sigma points can capture are not limited to the first two moments because UT offers flexibility to allow more information than mean and covariance to be incorporated. The algorithms that capture the first four moments of a Gaussian distribution and the first three moments of an arbitrary distribution have been presented in [8] [9].

4 Case Studies

To evaluate the performance of the presented EKF and UKF, both estimation algorithms have been applied to a nonisothermal continuous stirred tank reactor (CSTR). The CSTR is considered with coolant jacket dynamics, where the following exothermic irreversible reaction between sodium thiosulfate and hydrogen peroxide is taking place [10]:

$$2Na_2S_2O_3 + 4H_2O_2 \rightarrow Na_2S_3O_6 + Na_2SO_4 + 4H_2O \tag{16}$$

The capital letters A and B are used to denote the chemical compounds $Na_2S_2O_3$ and H_2O_2. The reaction kinetic law is reported in the literature to be [11].

$$\begin{aligned} -r_A &= k(T)C_AC_B \\ &= k_oe^{-E/RT}C_AC_B \end{aligned}$$

A mole balance for species A and energy balances for the reactor and the cooling jacket result in the following nonlinear process model:

$$\begin{aligned} \frac{dC_A}{dt} &= \frac{F}{V}(C_{Ain} - C_A) - 2k(T)C_A^2 \\ \frac{dT}{dt} &= \frac{F}{V}(T_{in} - T) + 2\frac{(-\Delta H)_R}{\rho c_p}k(T)C_A^2 - \frac{UA}{V\rho c_p}(T - T_j) \\ \frac{dT_j}{dt} &= \frac{F_w}{V}(T_{jin} - T_j) + \frac{UA}{V_w\rho_w c_{pw}}(T - T_j) \end{aligned} \tag{17}$$

where F is the feed flow rate, V is the volume of the reactor, C_{Ain} is the inlet feed concentration, T_{in} is the inlet feed temperature, F_w is the feed flow rate of the cooling jacket, V_w is the volume of the cooling jacket, T_{jin} is the inlet coolant temperature, c_p is the heat capacity of the reacting mixture, c_{pw} is the heat capacity of the coolant, ρ is the density of the reaction mixture, ρ_{pw} is the density of the coolant, U is the overall heat-transfer coefficient, and A is the area over which the heat is transferred. The process parameter values are given in [10].

4.1 State Estimation

The nonlinear model without process noise exhibits multiple steady states, of which the upper steady state (i.e. $C_{Ass} = 0.019$mol/L; $T_{ss} = 384.0\ K$; $T_{js} = 371.3\ K$) is stable and chosen as the point of operation.

In this case study, it is assumed that the system and measurement model is of the form

$$\begin{aligned} x_k &= f(x_{k-1}, u_{k-1}, k-1) + Gw_{k-1} \\ y_k &= Cx_k + v_k \end{aligned} \tag{18}$$

where $x_k = \begin{bmatrix} C_A & T & T_j \end{bmatrix}^T$, $w_{k-1} \sim \mathcal{N}(0, Q_{k-1})$ and $v_k \sim \mathcal{N}(0, R_k)$ are zero mean Gaussian noise. It is further assumed that only the third state Tj is measured. It is also assumed that process noise is present in the model with a magnitude of [0.00025 mol/l; 0.5 K; 0.5 K] and measurement noise of 0.5 K.

The EKF and UKF filter parameters are

$$\hat{x}_0 = \begin{bmatrix} 0.018 & 382 & 371.3 \end{bmatrix}^T, \qquad R_0 = 2.5 \quad \text{and} \quad \kappa = -4,$$
$$\hat{P}_0 = \begin{bmatrix} 6.25 \times 10^{-8} & 0 & 0 \\ 0 & 0.25 & 0 \\ 0 & 0 & 0.25 \end{bmatrix}, \quad Q_0 = \begin{bmatrix} 6.25 \times 10^{-9} & 0 & 0 \\ 0 & 0.025 & 0 \\ 0 & 0 & 0.025 \end{bmatrix}. \tag{19}$$

κ is not further adjusted to improve UKF performance so that the comparison of EKF and UKF is made purely on their mechanism.

The nonlinear system behavior and the performance of the extended Kalman filter and the unscented Kalman filter are shown in Figure 1, where the performance of linearized Kalman filter (LKF) is also illustrated for comparison purposes. In the case study it is assumed that the initial

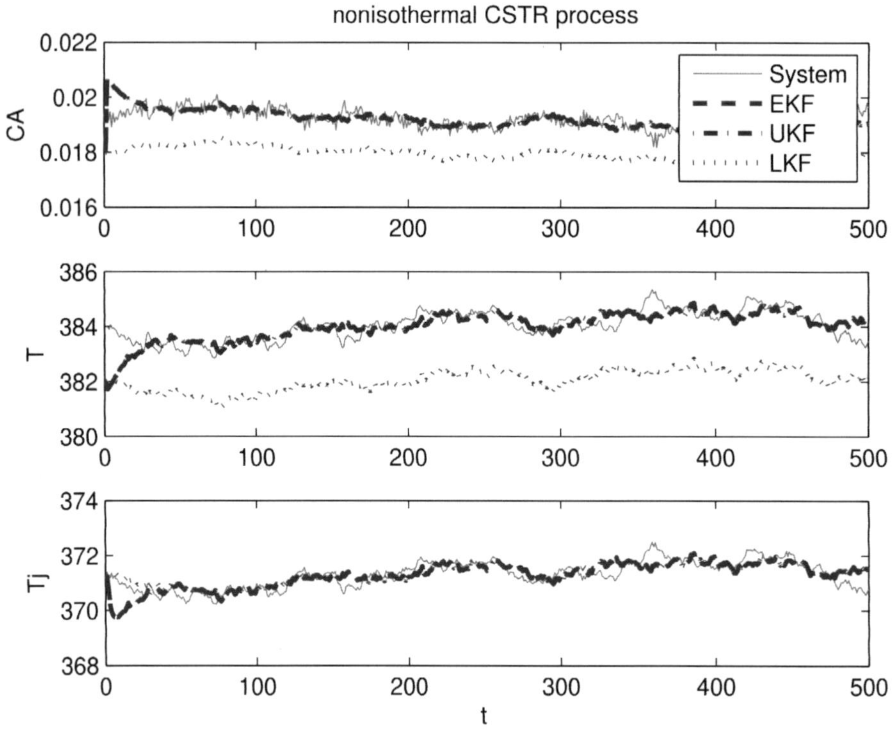

Figure 1: Comparison of Performance of each filter.

conditions of the unmeasured states are not precisely known. Therefore, there are offsets for the first two states by using LKF. EKF performs comparably to UKF because the given CSTR is operated around the upper steady state, which is in a locally linear region. Figure 2 illustrates the system behavior with the input changes. The dash-dotted line refers to the upper steady state where inlet coolant flowrate Fw is 30 L/min while the dotted line refers to the unstable steady state where $Fw = 225L/min$.

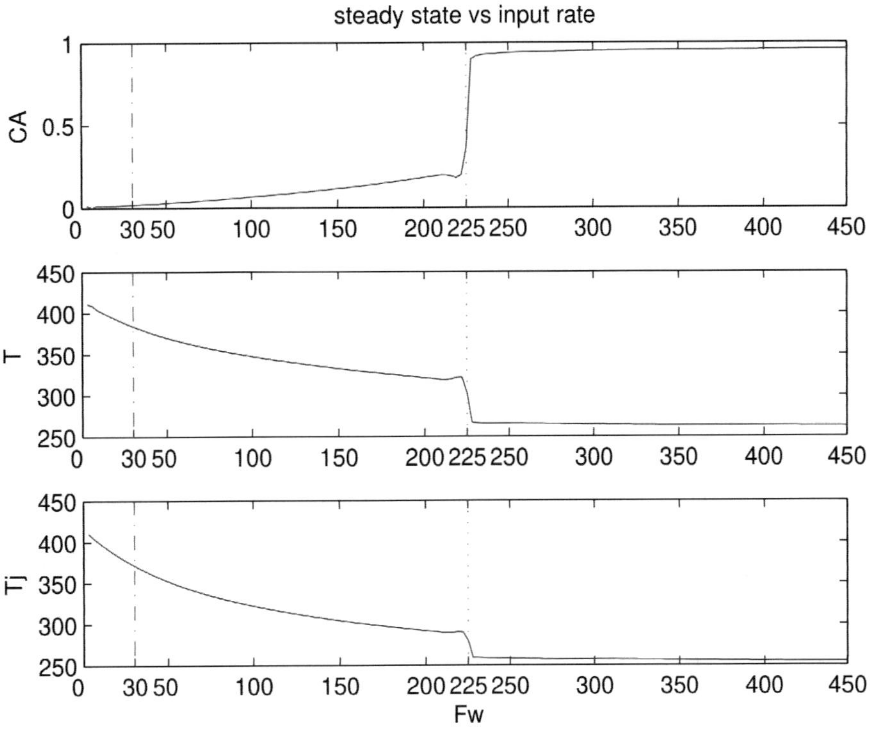

Figure 2: steady state vs input rate.

Figure 3 shows the results when the input change is doubled and applied to the nonlinear system. The system is driven to another steady state due to this change in the input. The results show that both the extended Kalman filter and the unscented Kalman filter perform well in estimating system states and outperform the linearized Kalman filter.

4.2 Parameter Estimation

As an extension of the UKF applications, UKF is implemented for parameter estimation in this section and results are compared to estimation with LKF and EKF.

In the case study $UA = 20000 * 60J/s.K$ is considered the parameter to be estimated since a process fault such as heat exchanger fouling can be detected on time by monitoring heat transfer

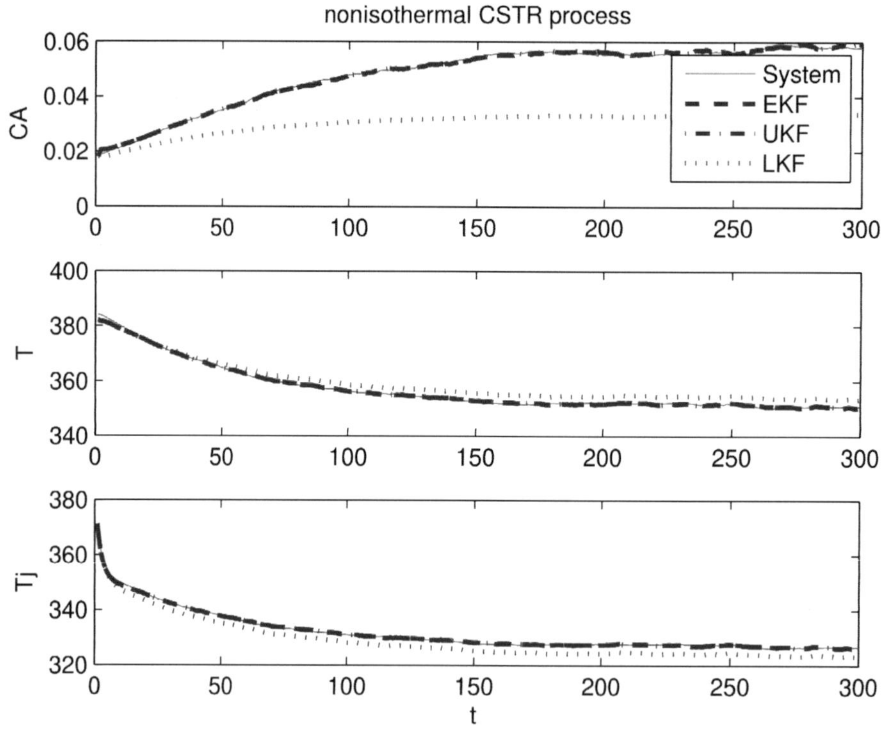

Figure 3: Performance after input change is doubled.

coefficients. The process noise implemented on the heat transfer coefficient is assumed to be 1.7%. The initial states and the rest of filter parameters are the same as shown in Eq. (19).

The performance of LKF, EKF and UKF for estimating the heat transfer coefficient is illustrated in Figure 4. The system is operated in a locally linear region while at time = 100 there is a sudden change in heat transfer coefficient UA. Both EKF and UKF outperform LKF again in estimating states and parameters. The difference in the EKF and the UKF results is minor because the system behaves close to a linear system. Figure 5 is the enlarged picture of the performance of each filter on parameter estimation.

Additional simulations were performed where the inlet coolant flow rate was changed to 225 L/min. For these experiments UKF performed better than EKF. However, since this is a drastic change of operating conditions, the results are not shown here.

In addition to the case study presented here, a Van de Vusse model [12] and a pH system [4] have been investigated for a comparison of performance of EKF and UKF. Extensive results show that both UKF and EKF provide good estimates for state and parameter estimation.

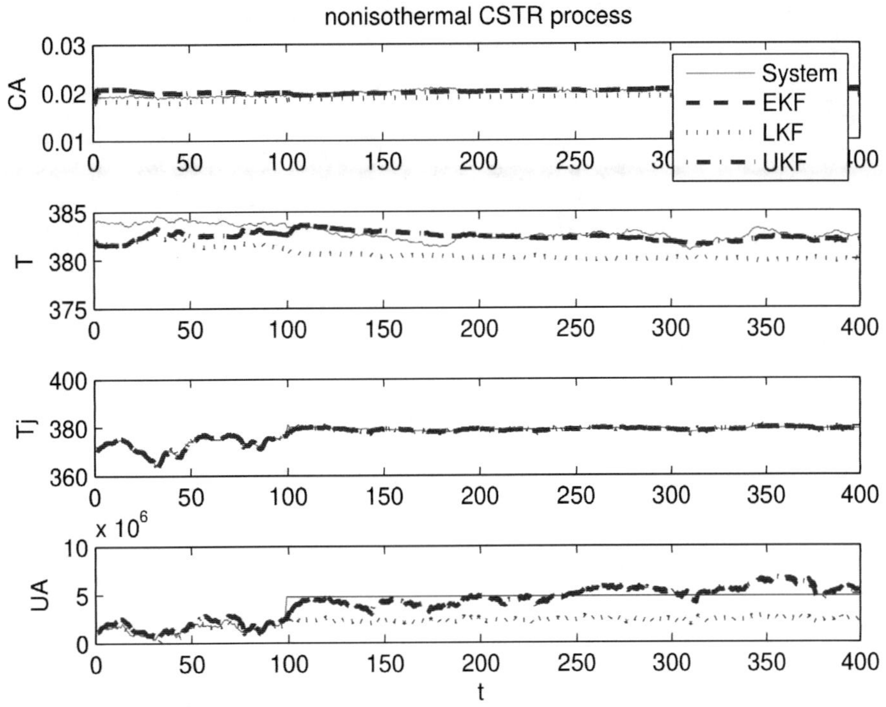

Figure 4: Performance for both states and parameter estimation.

5 Conclusions

The presented work investigated a nonlinear process monitoring approach via the unscented Kalman filter. The simulation results show that the unscented Kalman filter provides a good estimate of the states and parameters and is comparable in performance to EKF.

References

[1] S. J. Julier and J. K. Uhlmann. Unscented filtering and nonlinear estimation. *Proceedings of The IEEE*, 92(3):401–422, 2004.

[2] J. B. Rawlings and B R. Bakshi. Particle filtering and moving horizon estimation. *Computers and Chemical Engineering*, 30:1529–1541, 2006.

[3] A. Romanenko, L. O. Santos, and P. Afonso. Unscented kalman filtering of a simulated ph system. *Ind. Eng. Chem. Res.*, 43:7531–7538, 2004.

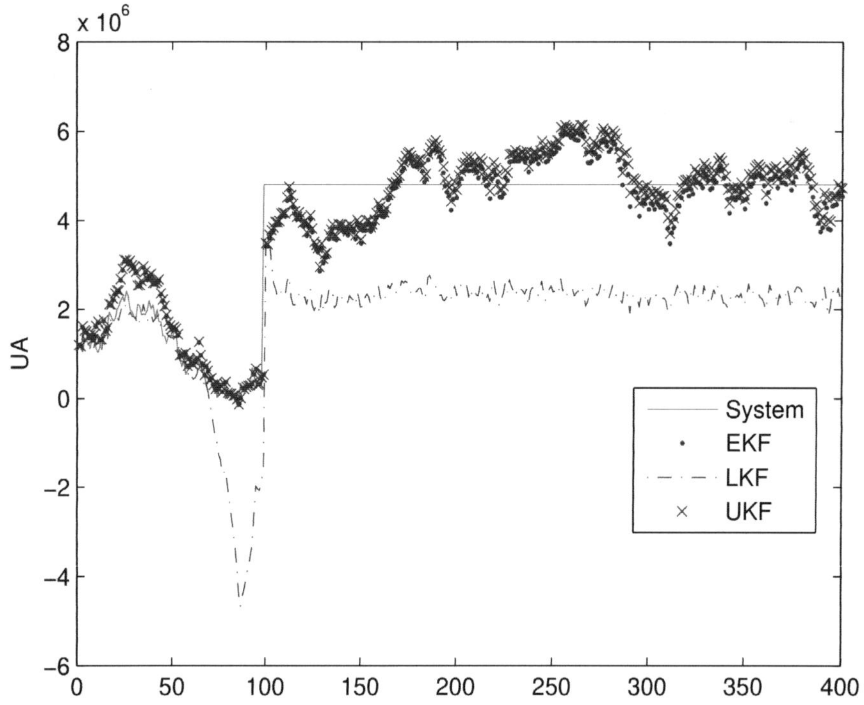

Figure 5: Performance for parameter estimation.

[4] A. Romanenko. and J. A.A.M. Castro. The unscented filter as an alternative to the ekf for nonlinear state estimation: a simulation case study. *Computers and Chemcial Engineering*, 28:347355, 2004.

[5] S. Rajaraman, J. Hahn, U. Kruger, and M.S. Mannan. Sensor fault diagnosis for time-varying dynamic systems using subspace model identification. 2006.

[6] M. S. Grewal and A. P. Andrews. *Kalman filtering : theory and practice.* Prentice-Hall, Englewood Cliffs, N.J., 1993.

[7] G. T. Schmidt. Practical aspects of kalman filtering implementation. *NATO Advisory Group for Aerospace Research and Development*, 1974.

[8] S. J. Julier and J. K. Uhlmann. A new extension of the kalman filter to nonlinear systems. *In Proc. AeroSense: 11th Int. Symp. Aerospace/Defense Sensing, Simul. and Controls*, pages 182–193, 1997.

[9] S. J. Julier. A skewed approach to filtering. *In Proc. AeroSense: 12th Int. Symp. Aerospace/Defense Sensing, Simul. and Controls*, 3373:54–65, 1998.

[10] S. Rajaraman, J. Hahn, and M.S. Mannan. A Methodology for Fault Detection, Isolation and Identification for Nonlinear Processes with Parametric Uncertainties. *Ind. Eng. Chem. Res.*, 43(21):6774–6786, 2004.

[11] S. A. Vejtasa and R. A. Schmitz. An experimental study of steady-state multiplicity and stability in an adiabatic stirred reactor. *AIChE J*, 16(3):410–419, 1970.

[12] J. Hahn and T. F. Edgar. A gramian based approach to nonlinearity quantification and model classification. *Ind. Eng. Chem. Res.*, 40:5724–5731, 2001.

DETECTING CHANGE IN COMPLEX PROCESS SYSTEMS WITH PHASE SPACE METHODS

C. Aldrich and G.T. Jemwa*

Department of Process Engineering, University of Stellenbosch,
Private Bag X1, Matieland, 7602, South Africa, Fax +27(21)8082059,
*E-mail: ca1@sun.ac.za

Introduction

Fault detection and identification in complex processes are particularly challenging, since the nonlinear behaviour of these processes may be difficult to model sufficiently accurately to detect anomalous conditions timeously. For example, feedback control loops containing sticking valves may generate self-sustained and limit cycle oscillations that can be propagated plantwide through physical coupling and recycles [1,2]. The source of these oscillations may be difficult to locate with conventional techniques. Likewise, in corrosion systems, a change in the nature of the process may result in changes in current and potential signals that may be difficult to detect with linear methods [3-6].

The use of phase space methods to detect change in nonlinear systems, such as those mentioned above, has only recently started to attract attention in various research communities, including chemical engineering (e.g. [7-10]). In this extended abstract, methods to detect change in nonlinear systems are briefly considered, including an approach based on the use of empirical orthogonal functions or singular spectrum analysis, a method based on the correlation dimension statistic of the time series data, and residuals generated by a model fitted to the data. All these methods are phase space methods, in the sense that their implementation rely on the embedding of the time series data in a phase space, as is explained in more detail below.

Phase Space Reconstruction

Phase space analysis of time series data starts with the embedding of the data in a so-called phase space (also referred to as state space or pseudostate space). In simple terms, this means that values of the time series, is transformed into a matrix, the columns of which contain the time series, as well as lagged copies of this time series. The embedding depends on two parameters, viz. a lag, k, and an embedding dimension m, as illustrated in Figure 1.

This results in an m-dimensional attractor with some geometrical structure, depending on the frequencies with which different parts of the attractor are visited. The features of this attractor or trajectory that can be visualized in two or three dimensions in the form of a so-called lag plot can yield important clues as to the nature and behaviour of the system. In essence, two systems can be considered identical, if their attractors are identical or at least if the two attractors have identical topologies.

Different approaches to the reconstruction of attractors from time series data are possible, The method of delays is a popular approach to the reconstruction of attractors from time series data, and finding the optimal lags (k) and dimensions (m) for embedding the time series is briefly outlined below.

Lag or Delay Time

Different lags (k) lead to reconstructions that are diffeomorphically equivalent, but geometrically different. If the lags are too small, the coordinates will not differ much from each other, resulting in state points scattered along the diagonal, and incomplete unfolding of the attractor. This *redundance* makes the estimation of the dimensionality difficult, because of the very small scales that have to be investigated, the presence of noise, etc. On the other hand, too large a time lag results in uncorrelated data to be included, which can cause an attractor to have a false complex shape [11]. In this investigation, the optimal lag k, was defined as the lag where the autocorrelation function of the time series reached a first minimum value.

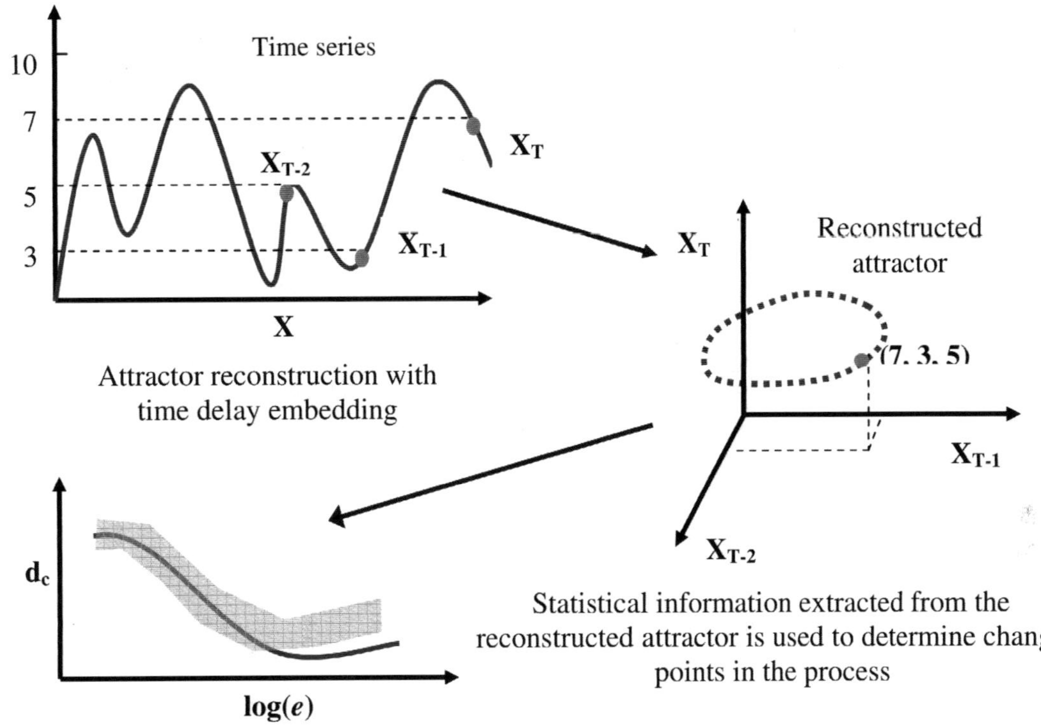

Figure 1. Embedding of an imaginary time series with a unity lag, k = 1, and an embedding dimension m = 3. The reconstructed attractor on the right is a graphical representation of the system dynamics. Statistical criteria derived from the attractor are used to monitor changes in the system.

Embedding Dimension

The optimal choice of the embedding dimension m is the smallest value that provides a proper reconstruction of the attractor. Too large an embedding dimension (m) results in very thin distributions of data in the reconstructed space, while too small an embedding dimension will lead to improperly reconstructed state spaces. With the false nearest neighbours algorithm [12], neighbouring observations or points are identified by unfolding the attractor in space of increasing dimension, and eliminating spurious neighbouring points discovered in the process.

More formally, the relationship between any given point on an attractor, y_j, and its potential nearest neighbour, y_j^{NN} in a state or phase space of dimension *d*, is examined by first adding another coordinate to the state space. The proximity of the two points in the state space is measured by the Euclidean distance between them. In an $\Re^d$-embedding, each point

$$y_j = [s_j, s_{j+k}, s_{j+2k}, \ldots s_{j+(d-1)k}] \quad (1)$$

has a nearest neighbour

$$y_j^{NN} = [s_j^{NN}, s_{j+k}^{NN}, s_{j+2k}^{NN}, \ldots s_{j+(d-1)k}^{NN}] \quad (2)$$

With a large number of observations, the distance between them is relatively small. Expanding the state space to $\Re^{d+1}$ yields the same points,

$$y_j^* = [s_j, s_{j+k}, s_{j+2k}, \ldots s_{j+(d)k}] \tag{3}$$

$$y_j^{*NN} = [s_j^{NN}, s_{j+k}^{NN}, s_{j+2k}^{NN}, \ldots s_{j+(d)k}^{NN}] \tag{4}$$

in a $(d+1)$-dimensional space. If the points are true neighbours, the distance between them (based on the last coordinates only)

$$\| y_j^* - y_j^{*NN} \|^2 - \| y_j - y_j^{NN} \|^2 = (s_{j+(d)k} - s_{j+(d)k}^{NN})^2 \tag{5}$$

should increase relatively slowly with successive increases in the embedding dimension. The normalized increase in distance between these two points is usually calculated and the points are considered to be false nearest neighbours, if

$$|s_{j+(d)k} - s_{j+(d)k}^{NN}| / \| y_j - y_j^{NN} \| \geq R_T \tag{6}$$

where a suitable value for R_T depends on the spatial distribution of the embedded data (typically $10 \leq R_T \leq 30$ [13]). Embedding of the time series data by means of the method of delays, as briefly summarized above, results in a trajectory of the dynamic system in a so-called phase space. In deterministic systems, with relatively little measurement noise, this trajectory tends to form a smooth orbit or attractor, the topology or shape of which is characteristic of the dynamics of the system.

Simply viewing the attractors in two or three dimensions often convey useful information, not readily apparent from direct observation of the time series data. However, for more formal detection schemes, the shape or topology of the attractor can be characterized by a number of useful statistics, chief of which would be various measures of the dimensionality of the attractor [14]. The correlation dimension of the time series is often used to this end.

Correlation Dimension as Test Statistic to Detect Change

The correlation integrals (defined by equation 7) are estimated from

$$\hat{C}_m(\varepsilon) = \binom{n}{2}^{-1} \sum_{i=1}^{n-1} \sum_{j=i+1}^{n} \theta\left(\varepsilon - \|v_i - v_j\|\right) \tag{7}$$

where n is the number of points in the data set. The maximum and minimum value the correlation function can attain are $C(\varepsilon) = 1$ and $C(\varepsilon) = 2/[n(n-1)]$ respectively. The idea is illustrated in Figure 2(a), which shows a hypersphere of radius ε traversing all the points on the attractor, counting all neighbouring points covered by the volume of the hypersphere.

A natural approach towards the characterization of the dynamics of a system is based on the characteristics that are invariant under reconstruction of the attractor of the system. The correlation dimension in particular is a popular measure and can be estimated by first determining a correlation sum $C(\varepsilon)$. The idea is illustrated in Figure 2. It consists of centering a hypersphere of radius ε on each successive point on the attractor and counting all the points within this hyperspheric region. The average sum of points within the hypersphere depends on the size of the hypersphere and the correlation dimension is a measure of this scaling factor. More formally, the calculations can be summarized as follows [15],

$$\hat{C}_m(\varepsilon) = \binom{n}{2}^{-1} \sum_{i=1}^{n-1} \sum_{j=i+1}^{n} \theta\left(\varepsilon - \left\| v_i - v_j \right\|\right) \quad (8)$$

where $\theta(\cdot)$ denotes the Heaviside function, i.e. $\theta(w) = 0$, for $w < 1$ and $\theta(w) = 1$, for $w \geq 1$ and $\|\cdot\|$ is some norm, such as the supremum norm $\|(x_1, x_2, \ldots x_m)^T\| = \sup_i|x_i| = \max(|x_1|, |x_2|, \ldots |x_m|)$ or the Euclidean norm. n is the number of data points in the time series. v_i is the data point on which the hypersphere is centred and v_j is any other point being considered. The maximum and minimum values the correlation function can attain are $C(\varepsilon) = 1$ and $C(\varepsilon) = 2/[n(n-1)]$ respectively. Interested readers can consult Addison [14], among other, for more detail.

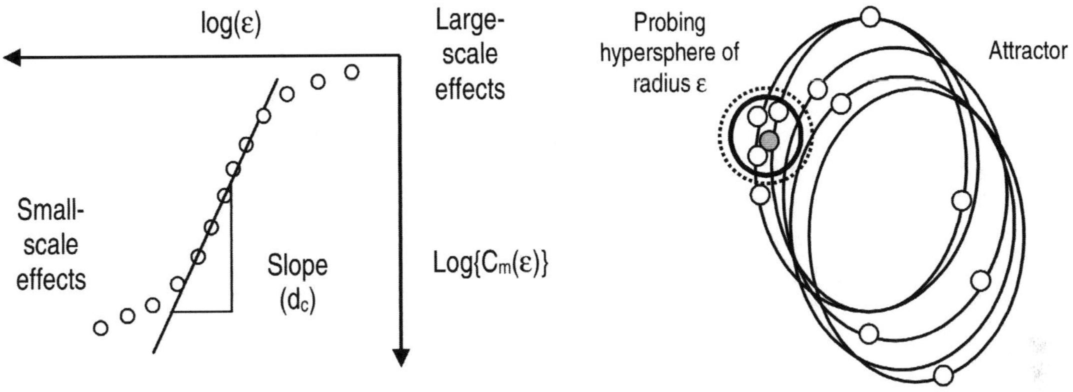

Figure 2. (a) Calculation of correlation sums by use of a probing hypersphere of radius ε moving along the attractor, and (b) calculation of the correlation dimension (d_c), as the derivative of the functional relationship between the correlation sum, $C_m(\varepsilon)$ and the scale (ε).

For deterministic time series with large m and small ε, the correlation integrals scale according to

$$\log\{C_m(\varepsilon)\} \approx d_c \log(\varepsilon) - K \quad (9)$$

where K is some constant. The next step is to estimate the correlation dimension from the sample correlation integral (equation 8). Strictly speaking, the scaling law only holds in the limit as $\varepsilon \to 0$, and $n \to \infty$, that is

$$D_2 = \lim_{\varepsilon \to 0} \lim_{n \to \infty} \frac{\log C_m}{\log \varepsilon} \quad (10)$$

This is illustrated in Figure 2(b). The correlation dimension is an invariant measure of the topological structure of an attractor that loses its invariant property when the system dynamics change. In this study, the correlation dimension was estimated over a range of scales by the Judd algorithm [16].

Methodology for Detection of Change

The methodology for detecting change in complex dynamic systems can be formalized as follows

i. Collect the time series data representative of normal process behaviour

ii. Determine a suitable lag and dimension (i.e. determine the lag coinciding with the first minimum in the estimated autocorrelation function of the time series and determine the embedding dimension with the method of false nearest neighbours).

iii. Divide the data into a number of segments of length R and embed each segment using the above parameters (lag k and embedding dimension m).

iv. Estimate the correlation dimension curves of the segments of the normal data as a function of scale. From these data, estimate confidence limits for the $d_c(\varepsilon_0)$-curve characterizing the topology of the time series attractor (dynamics of the system).

For monitoring of the system

v. Collect new process data.

vi. When sufficient data (L samples) have been collected, embed the data with the same parameters (lag and embedding dimension) calculated previously.

Determine the correlation dimension of these data.

vii. If the correlation dimension curve falls outside the confidence limits of the reference data, signal a change in the dynamics of the system.

The approach is demonstrated in the following case study.

Case Study: Autocatalytic Process

The autocatalytic process considered in this simulation consists of two parallel, isothermal autocatalytic reactions taking place in a continuous stirred tank reactor (CSTR) [17]. The system is capable of producing self-sustained oscillations based on cubic autocatalysis with catalyst decay at certain parameters.

The ordinary differential equations governing the system are subsequently defined by

$$\frac{dX}{dt} = 1 - X - aXZ^2 \tag{11}$$

$$\frac{dY}{dt} = 1 - Y - bYZ^2 \tag{12}$$

$$\frac{dZ}{dt} = 1 - (1+c)Z + \gamma_1 aXZ^2 + \gamma_2 bYZ^2 \tag{13}$$

Lee and Chang [17] have shown that for $a = 18000$; $b = 400$; $c = 80$; $\gamma_1 = 1.5$; $\gamma_2 = 4.2$ and with initial conditions $X_0 = 0$; $Y_0 = 0$; $Z_0 = 0$, the system exhibits chaotic behaviour.

The system (8)-(10) was simulated with the ODE45 subroutine in MATLAB 7.2 at a sampling rate of 0.005. For the first 10 000 observations the same parameters as specified by Lee and Chang [17] were used. A simulated fault condition was then introduced through a slight increase in the feed concentrations of A and D. This caused an increase in the parameters γ_1 and γ_2. The parameters γ_1 and γ_2 were allowed to drift slowly from their starting values, $\gamma_1 = 1.5$ and $\gamma_2 = 4.2$, to $\gamma_1 = 1.55$ and $\gamma_2 = 4.25$, after which the parameters were again kept constant at their new values, $\gamma_1 = 1.55$ and $\gamma_2 = 4.25$, for another 10 000 observations.

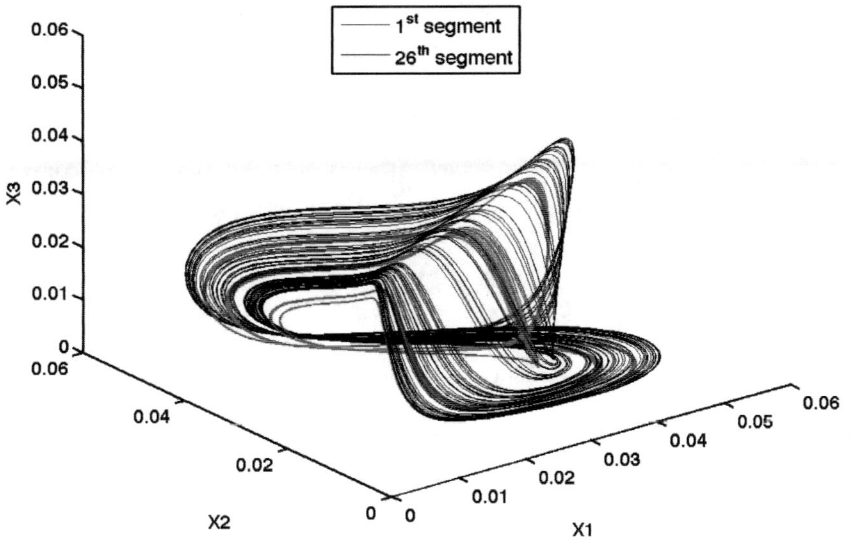

Figure 3. Some parts of 26th segment's attractor embedded in 1st segment's attractor for the autocatalytic data, embedded with a lag of 17 sample intervals.

The reconstructed attractor is shown in Figure 2, where the first 10 000 observations are indicated in black and the last 10 000 observations are indicated in red. As can be seen from this figure, it is difficult to distinguish the two stable states from one another from their attractors. However, the two states can easily be distinguished from each other from the correlation dimension curves shown in Figure 3. As before, the stable states are indicated by their mean curves and shading (95% confidence limits). The parameter drift from the initial stable state (shaded area at the top) is first signalled at 9000-14000, i.e. at some point between the 13 000th and 14 000th sample.

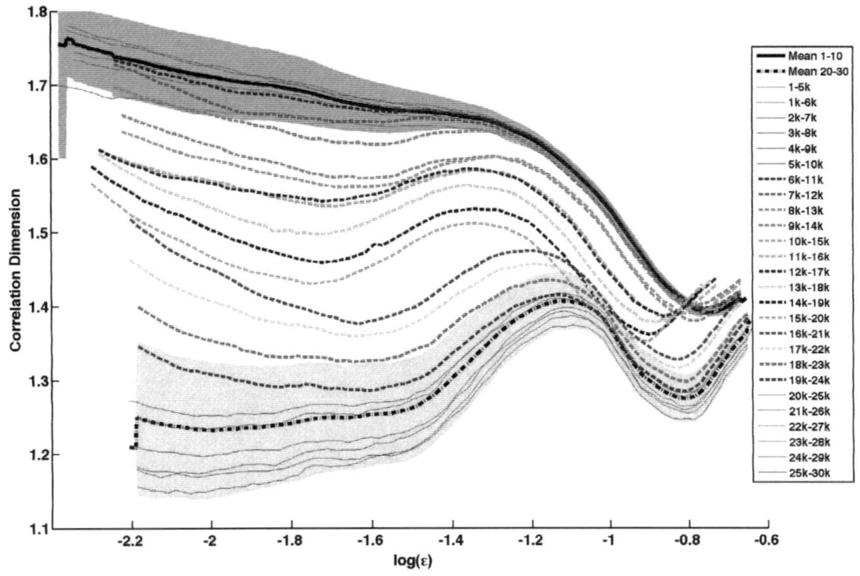

Figure 4. Correlation dimension estimates of the autocatalytic process data by the moving widow approach with the mean and 95% confidence intervals of the stationary parts.

The correlation dimension is a direct measure of the topology of the data, and should therefore be a sensitive measure of change in nonlinear systems, such as the ones considered in the case studies, provided that sufficient data are available for accurate estimates of the correlation dimension.

Comparison with Other Methods

The figures below show detection of change in the same data, with mutual cross prediction with multilayer perceptron neural networks, based on statistics characterizing the differences in the residuals from models fitted to the normal data, and then used to predict the test data, and vice versa. This approach was originally proposed by Schreiber [18], who has used a nearest neighbour model, instead of the neural networks considered here.

The test statistic $\gamma_{ij} = (S_i^r, S_j^r)$, the mean square prediction error, can be evaluated by:

$$\gamma_{ij} = (S_i^r, S_j^r) = MSE = \frac{1}{N - T(M-1) + 1} \sum_{t=T(d_e-1)+1}^{N} (y_t - \hat{y}_t)^2 \tag{14}$$

When diagonal segments are selected, the model is simulated with the data it was trained on and the test statistic $\gamma_{ij} = (S_i^r, S_j^r)$ should be expected to produce its lowest values.

A change in the dynamics of the systems is signalled when the prediction error is larger than a certain threshold, e.g. a 95% confidence limit on the errors associated with the segment on which the model was trained.

The results of a method based on the use of singular spectrum analysis, as proposed by Moskvina and Zhiglyavsky [19] is also shown in Figure 5 for comparison, where the data in a moving window are embedded and compared with subsequent segments of the embedded time series via a distance measure and a CUSUM statistic. In this case, the data were not embedded with the false nearest neighbour algorithm, but based on singular value decomposition of the lagged correlation matrix of the system.

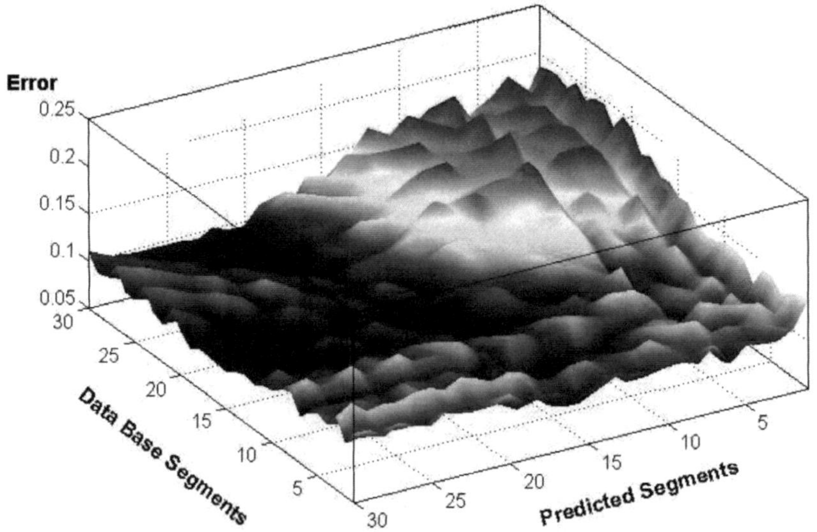

Figure 5. Mutual cross prediction of the autocatalysis data.

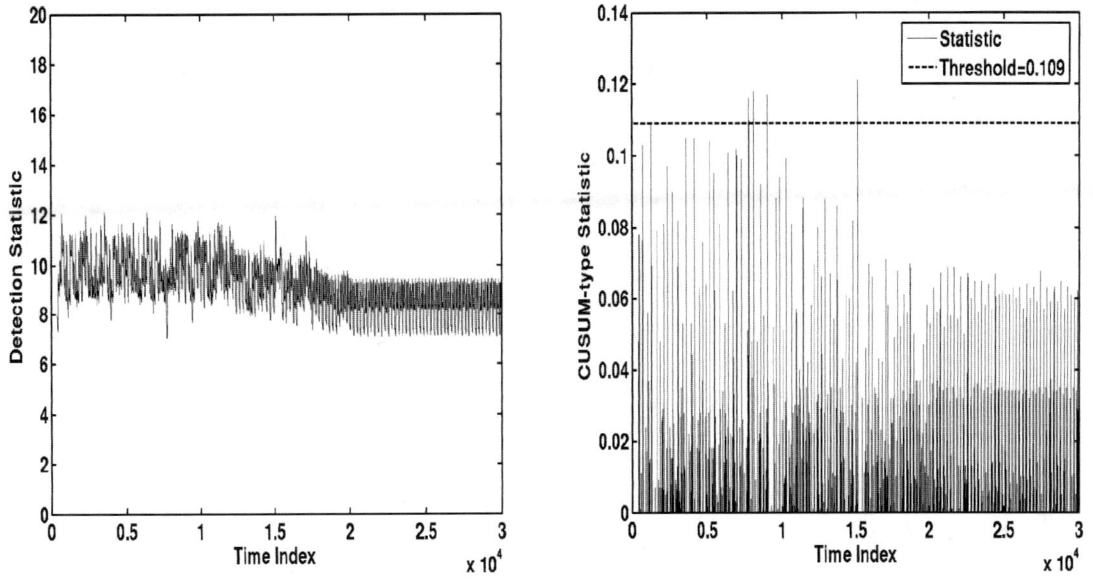

Figure 6. Summary of results from change-point detection algorithm for autocatalytic process data, showing the detection statistic $D_{n,I,p,q}$ ($m = 400$, $Q = 400$ and $p = 14$) (left) and the CUSUM statistic W_n (right).

The method based on the correlation dimension could detect the change within an average run length of 3000, as opposed to 5000 and 6000 for the mutual cross prediction method and that of based on singular spectrum analysis. Nonetheless, these conclusions cannot be generalized without more experimentation. Moreover, the method of singular spectrum analysis can easily be extended to a number of nonlinear variants. This work is currently underway, as is extension of these methods to various approaches dealing with multivariate time series data, as well as simulation of more realistic systems characterized by noisy measurement data.

Discussion and Conclusions

In this extended abstract, basic concepts in the use of phase space methods to detect change in complex dynamic systems are outlined and illustrated by means of a case study of an autocatalytic reaction in a continuous stirred tank reactor.

The results suggest that use of the correlation dimension, with bootstrapped estimates of confidence limits, is a promising approach that can be used to detect change in nonlinear systems, where other approaches are less successful.

Two other approaches proposed by other authors were also briefly considered. Both of these approaches appeared to be similarly promising. All the methods considered here are computationally intensive, among other owing to the bootstrap approach used to generate confidence limits (correlation dimension and mutual cross prediction), as well as the large number of calculations associated with the use of moving windows in all three the methods.

These phase space methods can be modified in a various ways. Since all of them depend on embedding of the data, different embedding strategies could conceivably offer some benefits not apparent at this stage. This includes the use of other criteria for determination of the embedding lag (such as average mutual information), or embedding of the data via nonlinear principal component analysis, for example.

Moreover, other types of dimensions or information criteria can be used in lieu of the correlation dimension (although preliminary work by the authors on the use of Lyaponuv features and information entropy [14] not reported here, did not seem to afford major advantages over the correlation dimension).

References

1. Åström, K.J. (1991). Assessment of achievable performance of simple feedback loops. International Journal of Adaptive Control and Signal Processing, 5, 3-19.

2. Shinskey, F.G. (2000). The three faces of control valves. *Control Engineering*, July.

3. Legat, A., Dolecek, V. (1995). Chaotic analysis of electrochemical noise measured on stainless steel. *Journal of the Electrochemical Society*, 142(6), 1851-1858.

4. Legat, A., Osredkar, J., Kuhar, V., Leban, M. (1998). Detection of various types of corrosion processes by the chaotic analysis of electrochemical noise. *Materials Science Forum*, 289-292 (PART 2), 807-812.

5. Garcia, E., Hernandez, M.A., Rodriquez, F.J., Genesca, J. and Boerio, F.J. (2003). Oscillation and chaos in pitting corrosion of steel. *Corrosion*, 59(1), 50-59.

6. Aldrich, C., Qi, B.C. and Botha, P.J. (2006). Analysis of electrochemical noise with phase space methods. *Minerals Engineering* 19(14), 1402-1409.

7. Briongos, J.V., Aragón, J.M. and Palancar, M.C. (2006). Phase space structure and multiresolution analysis of gas-solid fluidized bed hydrodynamics: Part I – The EMD approach, *Chemical Engineering Science*, 61(21), 6963-6980.

8. Montesino, O.M.E., Rolo, N.A. and Azoy, C.A. (2003). The attractor dimension applied to monitoring and surveillance in nuclear power plants. *Progress in Nuclear Energy*, 43(1-4), 389-395.

9. Van Ommen, J.R., Coppens, M.-O., van den Bleek, C.M. and Schouten, J.C. (2000). Early warning of agglomeration in fluidized beds by attractor comparison. *AIChE Journal*, 46(11), 2183-2197.

10. Dyaokwski, T., Mann, R., Wang, S.J. and Holden, P.J. (1996). Characterizing mixing processes using deterministic chaos theory. *Institution of Chemical Engineering Symposium Series*, 140, 131-142.

11. Sauer, T., Yorke, J.A., Casdagli, M. (1991). Embedology. *Journal of Statistical Physics*, 65, 579-616.

12. Kennel, M. B., Brown, R. and Abarbanel, H. D. I. (1992). Determining minimum embedding dimension using a geometrical construction. *Physica Review A,* 45, 3403-3411.

13. Small, M. and Judd, K. (1998). Correlation dimension: A pivotal statistic for non-constrained realizations composite hypothesis is surrogate data analysis. *Physica D*, Vol. 120, 386-400.

14. Addison, P.S. (1997). *Fractals and chaos.* London: IOP Publishing Ltd

15. Grassberger, P. and Procaccia I. (1983). Characterization of strange attractors. *Physical Review Letters,* 50, 346-349.

16. Judd, K. (1992). An improved estimator of the dimension and some comments on providing confidence intervals. *Physica D*, 56, 216-228.

17. Lee, J.S. and Chang, K.S. (1996). Applications of chaos and fractals in process systems engineering. *Journal of Process Control*, 6(2/3), 71-87.

18. Schreiber, T. (1996). Detecting and analyzing nonstationarity in time series using nonlinear cross predictions. *Physical Review Letters*, 78(5), 843-846.

19. Moskvina, V. and Zhigljavsky, A. (2003). Change-point detection algorithm based on the singular-spectrum analysis, *Communications in Statistics*, 32(2), 319-352.

Thermal Cracking Studies by Design of the Pilot Plant

Ehsan Bakhshi, R&D Center of National Petrochemical Company, Tehran, Iran and Fatemeh Abniki, Process Eng. Dept. of National Petrochemical Company,Ghadir Group, Tehran, Iran.

Thermal cracking of hydrocarbons from ethane up to gasoil is one of the most important processes for production of olefins, the basic feedstock for the petrochemical industry. The key to successful process control in the manufacture of ethylene, is temperature control of thermal cracking furnaces. This paper studies a computerized thermal cracking pilot plant and its capabilities for experimental activities. A computer program was developed for the monitoring and control purposes. The effect of temperature on ethylene yield is investigated by simulation and experimental studies. The results show good agreement between. Comparison of these results with industrial data, show that, they have the same trends. Coil Outlet Temperature (COT) is an important parameter affecting yield of ethylene production and therefore it should be controlled. Since furnaces are the first step in the production process, disturbances that occur due to the furnace operation will effect the entire process. COT is controlled by measuring the cracked gas temperature in the coil outlet and manipulating the heat input to the furnace. This loop is one of the most important loops in the control of thermal cracking furnaces. Fluctuations on COT is caused by the coke deposited on the COT measuring thermocouple. Furnace temperature tracks the furnace set point. Increasing the oscillation at higher set points, are due to nonlinear plant dynamic and using constant controller tunings. Based on the kinetic model a computer program for simulating the plant is developed. The software program is used to investigate the effects of temperature on the product yields. Simulation results indicate that increasing COT, increases the ethylene yield. Simulation results was tested experimentally by the pilot. Comparison of simulation, experimental and industrial results show, those have the same trends. The discrepancies between industrial and simulation results refer to the reactor configuration and the discrepancies between simulation and experimental results are due to the nonlinear and unknown furnace dynamic model, and some measurement errors in experiments.

Web Page: www.nipc.net

A Novel Mixed Product Run-to-Run Control Algorithm – Dynamic Ancova Approach

Ming-Da Ma[a], Chu-Cheng Chang[b], David Shan-Hill Wong[b] and Shi-Shang Jang[b,*]

[a]Center for Control and Guidance Technology, Harbin Institute of Technology, Harbin 150001, China

[b]Department of Chemical Engineering, National Tsing-Hua University, Hsin-Chu 30043, Taiwan

ssjang@mx.nthu.edu.tw **Abstract:**

In semi-conductor manufacturing industry, production resembles an automated assembly line in which many similar products with slightly different specifications are manufactured step-by-step, with each step being a complicated physiochemical batch process performed by a number of tools. This constitutes a high-mix production system for which effective run-to-run control (RtR) and fault detection control (FDC) can be carried out only if the states of different tool and different products can be estimated. However, since in each production run, a specific product is performed on a specific tool, absolute individual states of products and tools are not observable. In this work, a novel state estimation method based on analysis of variance (ANOVA) is developed to estimate the relative states of each product and tool to the grand average performance of this station in the fab. The method is formulated in a form of recursive state estimation using Kalman filter. The advantages of this method are demonstrated using simulations to show that the correct relative states can be estimated in production scenarios such as tool-shift, tool-drift, product ramp-up, tool-offline and preventive maintenance. Furthermore, application of this state estimation method in a minimum variance based RtR control scheme shows that substantial improvements in process capabilities can be gained, especially for products with small lot counts.

Keywords: state estimation, ANOVA, Kalman filter, run-to-run control

1. Introduction

The semi-conductor manufacturing industry is one of the fastest evolving industries in the world. As feature sizes shrink and wafer sizes increase, sophisticated control methods are needed to improve product yield, throughput, and overall equipment effectiveness.

The run-to-run (RtR) controller is a model-based process control system that integrates concepts in statistical quality control (SPC) and engineering process control (EPC). It is achieved by adjusting process inputs (recipes) at the beginning of each run based on information obtained from previous runs. In the last decade, RtR control has been extensively deployed in the semi-conductor industry. Research and development in this area have been summarized by many authors in books and review articles (Castillo and Hurwitz, 1997; Moyne, et al., 2001; Edgar, et al., 2000; Castillo, 2002, Qin, et al. 2006).

Most of RtR control algorithms are based on the assumption that there is only a single product fabricated in the manufacturing line (Butler & Stefani, 1994, Sachs, et al. 1995, Chen and Guo 2001). This is, however, far from reality. In semi-conductor manufacturing industry, production resembles an automated assembly line in which many similar products with different specifications are manufactured step-by-step, with each step being a complicated physiochemical batch process carried out by a number of tools. A specific combination of product and tool is known as a "thread" (Firth et al, 2006). Single product RtR control algorithms can be applied to a thread. However, the number of threads can be very large, upto thousands in a foundry fab. It is cumbersome to maintain so many controllers. Moreover, controller performance will be degraded for those infrequent threads since condition of the tool may be quite different from the last run of the same thread. Zheng et al. (2006) showed that even if the actual root cause is the change in condition of the tool, a single tool-based EWMA controller is unstable if the model uncertainties of different products are different. Pasadyn and Edgar (2005) and Firth et al. (2006) proposed to estimate contributions to biases states of each tool and product individually using previous runs and recombined them to determine recipe adjustments of future runs. However, in each production run, one product is manufactured on one tool. It can be shown that absolute individual states of products and tools are not observable. Pasadyn and Edgar (2005) include the qualifying wafers and assumed that the performance of qualifying wafer is equivalent to tool performance. However, there is always the dilemma of reduced throughout capacity if there are frequent runs of non-product test wafers and reduced sensitivity of the identification algorithm if only a few qualifying wafers are used. Firth et al. (2006) proposed a least square method known as just-in-time adaptive disturbance estimation (JADE) which includes additional constraints that the product and tool states remained unchanged from run-to-run and a proprietary weighting method.

In statistics, the problem of identification of different bias factors has been described as the analysis of variance (ANOVA) (Montgomery, 1997). In this work, a novel state estimation method based on ANOVA is developed to estimate the relative states of each product and the relative states of each tool to the grand average performance of this station in the fab. This method is formulated in a form of recursive state estimation using Kalman filter. The advantages of the proposed method are demonstrated using simulations to show that the correct ANOVA states can be estimated in production scenarios such as tool/product-shift, tool-drift, product ramp-up, product /tool-offline and preventive maintenance. Furthermore, application of this state estimation method in a minimum variance based RtR control scheme shows that substantial gains in process capability for specialized small lot counts products.

2. State estimation based on ANOVA

2.1 *Analysis of Variance* (*ANOVA*)

According to ANOVA, a multi-tool ($n=1,\ldots,N$) multi-product ($m=1,\ldots,M$) operation can be expressed as:

$$y_k - bu_k = \mu + \tau_{n_k} + p_{m_k} + \varepsilon_k \tag{1}$$

where b is the process gain which relate the change in manipulating variable u_k to the change in output quality variable y_k; μ is the overall mean of all observed tool and product combinations, τ_n ($n=1,\ldots,N$) represent the difference between the average results of all possible products on n^{th} tool and the overall mean, and p_m ($m=1,\ldots,M$) represent the difference between the average results on all possible tools of the mth product and the overall mean. ε_k is assumed to be white noise. τ_m and p_m are relative contributions subject to the constraints

$$\sum_{n=1}^{N}\tau_n = 0 \qquad \sum_{m=1}^{M} p_m = 0 \tag{2}$$

Here it is also assumed that there exist no interactions between tools τ_n ($n = 1,2,\cdots,N$) and parts p_m ($m = 1,2,\cdots,M$). In real practice, it should be noted that the output y is not only a function of present tool and product, but may also be influenced by other factors such as the specific tool used in the previous step. However, to include these factors is only a simple extension of this work. The matrix form of eq. (1) is

$$\left[\hat{Y}\right] = [Z][\alpha] = \begin{bmatrix} 1 & \delta_{1,n_1} & \cdots & \delta_{N,n_1} & \delta_{1,m_1} & \cdots & \delta_{M,m_1} \\ & & & \vdots & & & \\ 1 & \delta_{1,n_K} & \cdots & \delta_{N,n_K} & \delta_{1,m_K} & \cdots & \delta_{M,m_K} \end{bmatrix} \begin{bmatrix} \mu \\ \tau_1 \\ \vdots \\ \tau_N \\ p_1 \\ \vdots \\ p_M \end{bmatrix} \tag{3}$$

where $[Z]$ is an incidence matrix in which $\delta_{n,n_k}, \delta_{m,m_k}$ are Kronecker deltas, equal to *1* when n and m are equal to the tool n_k and product m_k of record k. For the incidence matrix $[Z]$, there are $N+M+1$ columns, but since

$$\sum_{j=2}^{N+1} Z_{ij} = 1 \qquad \sum_{j=N+2}^{N+M+1} Z_{ij} = 1 \tag{4}$$

the actual column rank is $N+M-1$, i.e. it is rank deficient by 2. However, given the two ANOVA constraints, we have

$$\left[\hat{Y}'\right] = \begin{bmatrix} \left[\hat{Y}\right] \\ 0 \\ 0 \end{bmatrix} = [Z'][\alpha] = \begin{bmatrix} & & [Z] & \\ 0 & \underbrace{1 \ \cdots \ 1}_{N} & \underbrace{0 \ \cdots \ 0}_{M} \\ 0 & \underbrace{0 \ \cdots \ 0}_{N} & \underbrace{1 \ \cdots \ 1}_{M} \end{bmatrix} [\alpha] \tag{5}$$

It should be pointed out that although the number of possible threads is NM, the possible number of different rows is greater than the number of independent columns. However, the actual number of threads is usually much less because some products with small lot counts may not be spread to all tools. There should be more than $N+M-1$ threads to make $[Z']$ full rank so that an unbiased estimate of $[\alpha]$ can be obtained. Furthermore, in some fabs, certain tools may be dedicated to produce certain products and vice versa. We must ensure that such groupings are treated as a separate system.

2.2 ANOVA State Space Model

If we assumed that the ANOVA states are stationary over several periods of time, then ANOVA model of the multi-tool and multi-product plant can be expressed in the following state space form:

$$[\alpha_{t+1}] = [T][\alpha_t] + [\omega_t] \tag{6}$$

$$\left[\hat{Y}'_t\right] = [Z'][\alpha_t] + [\upsilon_t] \tag{7}$$

where $[\omega_t]$ and $[\upsilon_t]$ are independent, zero-mean, Gaussian noise processes of covariance matrices $[Q]$ and $[R]$, respectively. $[T]$ is the transition matrix

$$T = \begin{bmatrix} 1 & 0_{1\times N} & 0_{1\times M} \\ 0_{N\times 1} & I_{N\times N} & 0_{N\times M} \\ 0_{M\times 1} & 0_{M\times N} & I_{M\times M} \end{bmatrix} \tag{8}$$

The system is observable if observability matrix for the above ANOVA state space model is

$$[O] = \left([Z'], [Z'][T] \cdots [Z'][T]^{N+M}\right)^T \tag{9}$$

The system is observable if the observability matrix is full rank. In this case, the transition matrix $[T]$ is an identity matrix, the observability matrix $[O]$ is equal to the output matrix $[Z']$ which is of full rank $N+M+1$.

2.3 Recursive Observation

Estimation can then be carried out in a recursive manner from intervals to intervals. At the start of any time interval t, given an estimated ANOVA state vector $[\tilde{\alpha}_{t-1}]$ and a estimated covariance matrix of the ANOVA states $[\tilde{P}_{t-1}]$, then the predicted values of the ANOVA state vector $[\tilde{\alpha}_{t|t-1}]$ and predicted the covariance matrix for this period $[\tilde{P}_{t|t-1}]$ are given by

$$[\tilde{\alpha}_{t|t-1}] = [T][\tilde{\alpha}_{t-1}] \tag{10}$$

$$[\tilde{P}_{t|t-1}] = [T][\tilde{P}_{t-1}][T]^T + [Q] \tag{11}$$

After the operating records $(y_{t,k}, u_{t,k}, n_{t,k}, m_{t,k})$, $k = 1 \cdots K_t$ of this period are collected, the minimum mean square estimator of the ANOVA states and the covariance matrix can be updated by the following equations (e.g., Duncan & Horn, 1972):

$$[\hat{\alpha}_t] = [\tilde{\alpha}_{t|t-1}] + [\tilde{P}_{t|t-1}][Z'_t]^T [\Phi_t]^{-1}\left([\hat{Y}'_t] - [Z'_t][\tilde{\alpha}_{t|t-1}]\right) \tag{12}$$

$$[\tilde{P}_t] = [\tilde{P}_{t|t-1}] - [\tilde{P}_{t|t-1}][Z'_t]^T [\Phi_t]^{-1}[Z'_t][\tilde{P}_{t|t-1}] \tag{13}$$

$$[\Phi_t] = [R] + [Z'_t][P_{t|t-1}][Z'_t]^T \tag{14}$$

In these circumstances a set of prediction error given by:

$$[v_t] = [\hat{Y}'_t] - [Z'_t][\tilde{\alpha}_{t|t-1}] \tag{15}$$

are produced by the Kalman filter. These are independently and normally distributed with mean zeros and variance $\sigma^2 \Phi_t$.

It is not possible to guarantee that $[Z'_t]$ contains all the threads and of full rank $N+M+1$ during the data collection interval. For example, in the extreme case, the model can be updated whenever the result of a single run is reported. However it is important to ensure that $[Z']$ over an extended history is of full rank $N+M+1$. This assumption may be invalid if some products are terminated or a certain tool is offline for an extended period of time. The assurance of the above full rank assumption can be achieved by monitoring the condition number of the matrix $\left[\tilde{P}_t\right]$.

2.4 Run-to-run control

The control objective is to maintain the process output as close to target as possible. Given a set of predicted states $\left[\tilde{\alpha}_{t|t-1}\right]$, the deadbeat control action in the period t, before new data arrive are given by:

$$u_{t,k} = \frac{T_k - \tilde{\mu}_t - \tilde{\tau}_{t,n_k} - \tilde{p}_{t,m_k}}{b} \tag{16}$$

3. Simulation study

In this section a series of simulation tests are designed to investigate the effective ness of the proposed algorithm in various operation scenarios. A simple two-tool-thr ee-product example is used in the following simulation studies. Note that it is easy to i dentify the changes of tools or products using the ANOVA estimator proposed in this paper. If the ANOVA states of the tools or products remain stable, the corresponding conditions of the tools or products are confirmed to be unchanged. In case of a chang e of the ANOVA state are observed, the change of $\tilde{\mu}+\tilde{\tau}_n$ or $\tilde{\mu}+\tilde{p}_m$ confirms the cha nge of the tool n or the product m. Furthermore, the changes of $\tilde{\mu}+\tilde{\tau}_n$ or $\tilde{\mu}+\tilde{p}_m$ are consistent with the actual states a_n^{tool} and a_m^{prod}.

3.1 Tool-shift

There are many events which can result in an apparent immediate shift in the operating conditions. For instance, such a disturbance might occur when a tool undergoes a maintenance event.

The comparisons of actual and estimated ANOVA states, the actual and estimated states of actual biases and relative bias of the each tool and products are shown in Figure 1. In this case, there is an abrupt change for the bias value of tool 1 at the 200th run. In Figure 1a, we found that the ANOVA states $\tilde{\mu}$ and $\tilde{\tau}_1$ experience positive shifts while the ANOVA tool state $\tilde{\tau}_2$ experience a shift in the other direction. In Figure 1b, it is found that the average performance of all products on tool 1 $\tilde{\mu}+\tilde{\tau}_1$ experienced a shift; the average performance $\tilde{\mu}+\tilde{\tau}_2$ of all products on tool 2 remained unchanged while the average performance $\tilde{\mu}+\tilde{p}_{A,B,C}$ of all products on tool 2 experienced a shift too.

3.2 Controller performance

In a high mixed foundry, some products are fabricated infrequently in small quantity. However, it should be noted that products which are produced with large quantity are usually of marginal profits and the products which are produced infrequently are often high-value added and contribute a substantial portion of the profit. Therefore, it is highly desirable for a control algorithm to have comparable performance for products with different run counts.

In this section, the performance for "infrequent" products for the three control algorithm, threaded EWMA algorithm, JADE and the ANOVA method proposed in this paper are investigated. Runs are evenly distributed between the two tools, and the probability distributions of products A, B, C are 60%, 35% and 5% respectively. Two IMA (1, 1) time series tool disturbances are injected into the system:

$$\eta_k = \eta_{k-1} + \varepsilon_k - \theta\, \varepsilon_{k-1} \tag{17}$$

where $\varepsilon \in N(0, \sigma^2)$ are a Gaussian distributed random noise with zero mean and variance $\sigma^2 = 0.16$. The mean squared errors (*MSE*) for different products are used as the performance index:

$$MSE_m = \frac{\sum_k \delta_{m,m_k} (T_k - y_k)^2}{\sum_k \delta_{m,m_k}} \tag{18}$$

The simulation results are shown in table 1. For all three methods, product A, which is the most frequent product produced, had the best performance. Product C, which is the produced in small quantity, had the worst performance. However, the performance of threaded EWMA is extremely poor for product C, while performances of JADE and ANOVA for product C are still acceptable. The ANOVA result is the best among the three methods for all three products with different run counts.

Table 1: Comparison of RtR control performances for different products

Method	MSE_A	MSE_B	MSE_C	MSE
ANOVA	0.2400	0.2833	0.4124	0.2642
JADE	0.2557	0.3182	0.5423	0.2964
EWMA(Threaded)	0.2746	0.4103	2.2348	0.4152

4. Conclusion

In this paper a novel state estimation method based on statistics method ANOVA is developed to estimate the relative states of each product and the relative states of each tool to the grand average of this station in the fab. The method is formulated in a form of recursive state estimation using Kalman filter. Simulation results show that the correct ANOVA states can be estimated in production scenarios such as tool-shift, tool-drift, product ramp-up and offline. Furthermore, application of this state estimation method in a minimum variance based RtR control scheme shows that substantial improvement of quality of products with small run counts. This makes the proposed method highly suitable for mixed product control system.

Acknowledgement

The authors thank the financial support provided by National Science Council through the grant NSC95-2221-E-007-027.

References

Bulter, S. W. & Stefani, J. A. (1994). Supervisory run-to-run control of polysilicon gate etch using in situ ellipsometry. *IEEE Trans. Semicond. Manuf.*, **7**, pp. 193-201.

Castillo, E. D., & Hurwitz, A. M. (1997). Run-to-Run Process Control: Literature Review and Extensions. *Journal of Quality Technology*, **29**, pp. 184-196.

Castillo, E. D. (2002). *Statistical Process Adjustment for Quality Control.* John-Wiley and Sons, New York.

Chen, A. & Guo, R. S. (2001). Age-based double EWMA controller and its application to CMP processes, *IEEE Trans. Semicond. Manuf.*, **14**, pp. 11–19.

Duncan, D. B. & Horn, S. D. (1972). Linear dynamic regression estimation from the viewpoint of regression analysis. *J. Am. Statist. Assoc.* **67**, 815-824.

Edgar, T. F., Butler, S. W., Campbell, W. J., Pfeiffer, C., Bode, C., Hwang, S. B., Balakrishnan, K. S., & Hahn, J. (2000). Automatic control in microelectronics manufacturing: Practices, challenges, and possibilities, *Automatica*, 36, no. 11, pp. 1567–1603, 2000.

Firth, S. K., Campbell, W. J., Toprac, A., & Edgar, T. F. (2006). Just-In-Time Adaptive Disturbance Estimation for Run-to-Run Control of Semiconductor Processes. *IEEE Trans. Semicond. Manuf.*, **19**, pp. 298-315.

Moyne, J. (2001). *Run-to-Run Control in Semiconductor Manufacturing*. CRC Press, Florida.

Montgomery, D. C. (1997). *Design and Analysis of Experiments*, 4th Ed. John Wiley&Sons, New York.

Pasadyn, A. J., & Edgar, T. F. (2005). Observability and State Estimation for Multiple Product Control in Semiconductor Manufacturing. *IEEE Trans. Semicond. Manuf.*, **18**, pp. 592-604.

Qin, S. J., Cherry, G., Good, R., Wang, J. & Harrison, C. A. (2006). Semiconductor manufacturing process control and monitoring: A fab-wide framework. *J. of Process Control*, **16**, pp. 179-191.

Sachs. E., Hu, A., & Ingolfsson, A. (1995). Run by Run Process Control: Combining SPC and Feedback Control. *IEEE Trans. Semicond. Manuf.*, **8**, pp. 26-43.

Zheng Y., Lin, Q. H., Wong, D. S. H., Jang, S. S., & Hui, K. (2006). Stability and Performance Analysis of Mixed Product Run-To-Run Control, *J. of Process Control*, **16**, pp. 431-443.

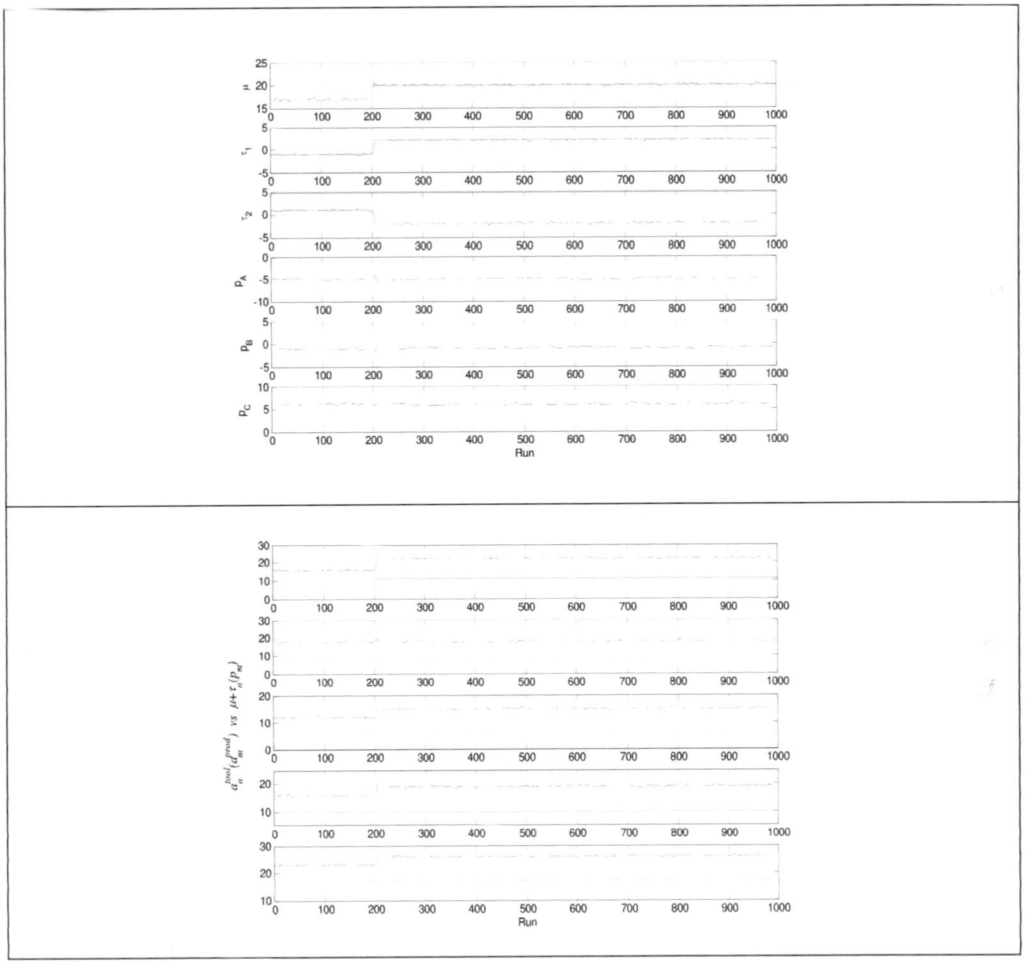

Figure 1: Changes in actual and estimated states of ANOVA parameters and factor biases in a tool shift. (— actual values, --- estimated values)

A Practical Approach to Advanced Process Control of VCM Plants

Ravi Nath, Mike Yen and Zak Alzein
Honeywell International, USA

Eric Wagner
Technip, USA

Abstract

Vinyl Chloride Monomer (VCM) is an industrially important chemical whose demand is growing at a fast pace worldwide. VCM is produced by thermal cracking of Ethylene Dichloride (EDC) which is a fouling service that gradually degrades the performance of cracking furnaces and therefore necessitates periodic shutdown and clean up of cracking furnaces.

Multi-variable Predictive Control (MPC) is a commonly applied technology to processing units to improve the stability, productivity and efficiency of the process. However, MPC cannot be readily applied to VCM plants because some of the key control variables, such as individual furnace EDC conversion, VCM production, coke thickness and tube skin temperature are difficult, if not impossible, to measure.

High fidelity EDC cracking yield model that accurately predicts EDC conversion, VCM production, coke buildup and tube skin temperature in EDC cracking furnaces does exist and could be used in an on-line fashion to provide continuous feedback to MPC. One such approach and its application to Advance Process Control of VCM plants are presented in this paper.

1. Introduction

Vinyl Chloride Monomer (VCM) is an industrially important bulk commodity chemical. It is primarily used to produce Poly Vinyl Chloride (PVC). PVC has fire retardant properties and is used extensively as building material in the form of pipes, conduits and insulation for electrical wiring. PVC is the second largest volume plastic in the world; worldwide production exceeds 15 billion pounds annually and is growing at a pace faster than the GDP.

2. Process Chemistry

Commercially VCM (C_2H_3Cl) is produced by gas phase thermal cracking of Ethylene Dichloride ($C_2H_4Cl_2$) which also produces Hydrogen Chloride (HCl) as a byproduct;

$$C_2H_4Cl_2 \rightarrow C_2H_3Cl + HCl$$

Typically about half of the EDC that is used is produced by direct chlorination of Ethylene

$$C_2H_4 + Cl_2 \rightarrow C_2H_4Cl_2$$

Whereas the other half of EDC that is used is produced by oxy-chlorination of Ethylene that consumes HCl that is produced by EDC cracking;

$$C_2H_4 + 2HCl + 1/2O_2 \rightarrow C_2H_4Cl_2 + H_2O$$

Overall the process is balanced and there is no net production of HCl and can be represented as;

$$2C_2H_4 + Cl_2 + 1/2O_2 \rightarrow 2C_2H_3Cl + H_2O$$

3. Process Description

The main process comprises the following 5 functional areas;

1. EDC production: to produce raw EDC by direct and oxy-chlorination.
2. EDC purification: to remove water and other impurities from raw EDC.
3. EDC cracking: to crack EDC that produces VCM and HCl, usually in multiple furnaces.
4. VCM purification: to purify VCM product from cracked gas by separating HCl and un-reacted EDC.
5. EDC recovery: to recover EDC from the aqueous phase prior to discharging to water treatment.

In addition there are support processes of refrigeration, water cooling, condensate recovery and incineration.

The figure below is a simplified process flow diagram.

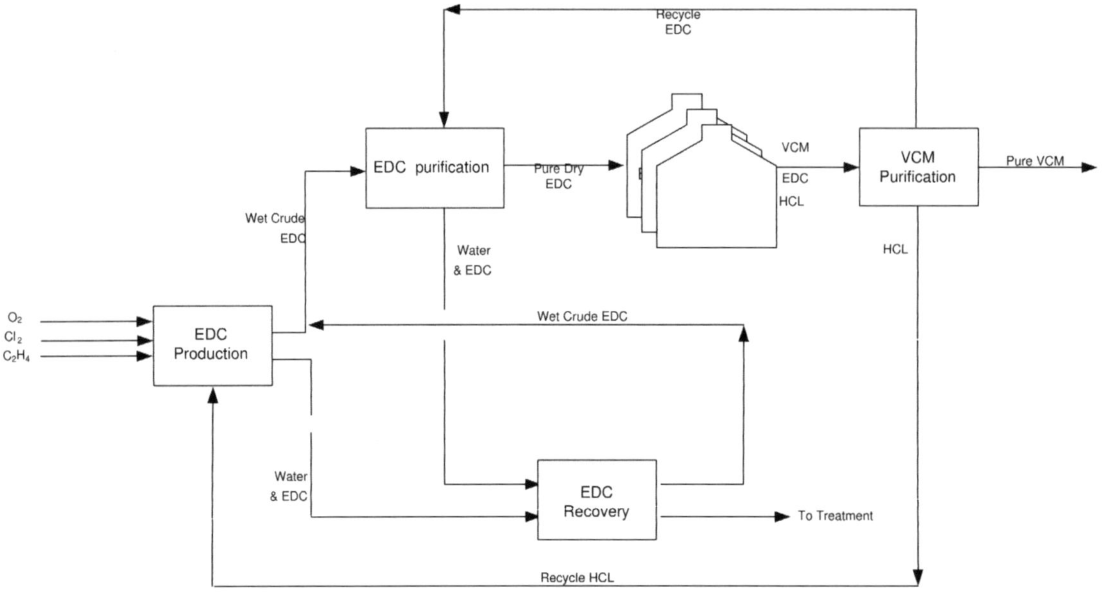

Fig 1 – VCM process schematic

4. Process Environment

The process is continually subjected to a variety of measured and unmeasured disturbances. Some of the disturbances are inherent to the process while others are due to external factors. Also, some of the disturbances are slow and gradual while others are sudden and abrupt.

Some disturbances are caused by the thermal cracking process itself. Thermal Cracking of EDC produces a small amount of solid coke material; this material deposits on the inside of the cracking tubes. Coke deposit on the cracking tubes increases the resistance to heat and momentum transfer which tend to reduce the amount of EDC cracking and VCM production. To compensate for this phenomena heat flux (fuel firing) is gradually increased which further increases coke production; ultimately the coke deposit on the tubes becomes so severe that the furnace has to be taken out of service for cleanup. The frequency of cleanup varies, but usually is once every 3 to 12 months. So while a furnace is in service, there is a gradual deterioration in it's performance. And when the furnace is taken out or brought back in service there is sudden large disturbance in the process. Other processing equipment such as heat exchangers and compressors also undergo similar gradual drift in performance.

Other disturbances are caused by external factors such as changes in the ambient conditions. Some of the external disturbances are gradual such as change in the ambient temperature while others are more sudden such as a rain storm or snow precipitation. In addition there could be gradual or sudden changes in the feed quality which may be measured but usually with some delay.

Needless to say, the process environment is constantly changing. The behavior of the process, especially thermal cracking is also non-linear which means linear prediction will at best be approximate. Additionally the process has dynamic aspect which means that the impact of a change is felt over a period of time, usually with some delay, by the down-stream process.

5. APC Technology

One of the objectives of process operations is to keep the process steady which keeps the product quality within specifications. This is a challenging task since the process is constantly subjected to numerous disturbances. A strategy to achieve this objective is by making counteracting moves to the variables that the process can manipulate. The counteracting moves have to be timely and anticipatory because response of the process to such moves is not immediate as it follows the process dynamics.

Traditionally, process control is achieved by a one to one paring between a controlled variable and a manipulated variable and by specifying a setpoint for the variable to be controlled. A continuous feedback from the process is required to monitor deviations from the setpoint. This is the so called PID algorithm. This Single Input Single Output (SISO) algorithm severs us well when there is little or no interaction between variables and when there is little or no process delay, an example being fluid flow control.

There are many circumstances where the simplistic PID type algorithm fails to deliver satisfactory process control. This happens when there are interactions among the process variables and / or there is significant process dynamics which is characterized by significant dead time and time constant. Fortunately, over the last two decades multivariable control technology has evolved and developed to an extent that control of all but the most difficult control problems can be achieved. Such Multi Input Multi Output (MIMO) technology, also called Multivariable Predict Control or Model Predictive Control (MPC) is commercially available through many technology vendors.

MPC relies on a dynamic model of the process, which is usually identified by plant testing. Similar to SISO, process feedback is an essential element of all MPC algorithms. Process feedback could be continuous such as temperature measurement or it could be an intermittent such from as from an analyzer or from a laboratory.

6. Use of Soft Sensor with APC

Difficulty arises when feedback from the process is impossible or not easily measurable. Examples abound in the cracking section. Coke thickness in the cracking furnace tube is simply not measurable with the current state of knowledge. VCM yield and EDC conversion from each furnace would require prohibitively expensive analyzers. And tube skin temperature although measurable is error prone.

So what can be done? A possible solution would be to use computations that could predict the un-measurable variables, this is the so called soft sensor approach. Fortunately, for the EDC cracking process such a model is commercially available. This model, described later in the paper, can accurately and reliably predict the coking rate, VCM and other component yields, EDC conversion, maximum tube skin temperature etc. The model is accurate and robust and can execute at a fairly high speed. By running this model with process data it would be possible to provide feedback to the furnace MPC controllers in a timely fashion.

7. APC Opportunities in VCM plant

In addition to the providing stability to the process by disturbance rejection, APC also offers possibility of production increase and energy usage reduction.

A benefits analysis is done to assess the benefits and identify APC projects with good payback for a particular plant. Possible APC applications with good payback include the following.

- Oxy-chlorination Reaction Section – Oxy-chlorination Reactor
- EDC Purification Section - Heads Column
- EDC Purification Section - Hi-Boil Column
- EDC Purification Section - Vacuum Column
- EDC Purification Section – Vacuum Bottom Flash Column
- EDC Cracking Section – Cracking Furnaces (multiple furnances, either liquid phase or vapor phase, depends on process)
- VCM Purification Section – HCL Columns (more than one, depends on the process)
- VCM Purification Section – VCM Column
- VCM Purification Section – VCM Stripper

Objectives of the APC applications include the following:

- Maximize Furnace throughput of the VCM plant subject to constraints
- Optimize Furnace run-length
- Minimize By-Product subject to constraints
- Minimize EDC/VCM loss subject to constraints
- Maximize VCM production subject to constraints
- Maintain product quality control
- Minimize energy consumption while trading off product recoveries

and capacity
- Enhance operability

MPC requires process feedback for all of the controlled variables (CVs). CVs for which direct measurement is not possible or cost effective, use of soft sensors (virtual sensors or inferential calculations) may be a workable strategy. Possible soft sensors in a VCM plant control include the following;

- Oxy Reactor: concentration of CCl_4 in the crude EDC Condenser.
- Oxy Reactor: component concentration of the vent gas stream.
- EDC purification - Heads Column: Concentration of CCl_4 in the column overhead and bottom streams.
- EDC purification - Hi-Boil Column: Concentration of CCl_4 in the column side product
- VCM column: concentration of combined impurities in the column overhead stream.
- EDC Cracking Furnace: VCM yield, EDC conversion, coking rate and tube skin temperature for each furnace pass.

8. Detailed description of an MPC

Honeywell's Profit Controller is a multivariable predictive controller technology that has been widely used in the refining and petrochemical industry. It has a unique design, the following are some of its important features [1];

Robust Controller Design: Profit Controller's patented design enhances the robustness of the on-line controller by explicitly taking into account the uncertainties associated with models used by the controller. This design gives the best control for the worst case, tolerating a considerable level of model error and maximizing controller uptime.

Range Control Algorithm: The patented Range Control Algorithm (RCA) is the core component of the Profit Controller. It controls all CVs (Controlled Variables) within their ranges, recognizing that a setpoint is simply a special case of a range. RCA enables the Profit Controller to explicitly constrain all of the CVs *dynamically* into the future. The number of CVs can be greater or fewer than the number of MVs (Manipulated Variables), output by RMPCT to achieve the CV control. RCA achieves the following:

- MVs are moved by the minimum amount required to achieve the control objectives.
- Almost-parallel CVs (CVs that have similar gains and dynamics) are controlled in a much more robust fashion than was possible with prior technology.
- Unification of constraint control, regulatory control, and process optimization is

inherent.

- Constraints are active only when they are limiting.
- Versatile control performance adapts to changing degrees of freedom when plant loads and constraints change.
- CVs may have integrating dynamics (e.g., tank levels).

RCA keeps Profit Controller online in situations where other multivariable controllers become unstable.

Product Value Optimization (PVO): A flexible (linear and/or quadratic) objective functions which user can specify optimization in terms of unit economics or other custom objectives. Economic factors can be changed online. PVO is dynamic rather than steady state solution, which lets Profit Controller optimize the entire unit faster and potentially offers a more precise final solution providing greater profitability.

PVO finds not only the steady-state solution, but also the optimal dynamic path to achieve that solution. PVO is treated as part of closed-loop control and is unified into the Range Control Algorithm. As a result, optimization speed can be tuned independently of the speed of CV error correction.

Advantages of Profit Controller

- Maximum process efficiency – Advanced algorithm balances performance and robustness objectives against process economics to minimize costly process movement.
- Flexibility to meet process needs – Configurable control response path allows tailoring of control performance to meet process needs.
- Optimum control performance – Independent feedforward and feedback control tuning provides optimum control performance for changes in both control targets and process disturbances.
- Enhanced robustness – Funnel based approach to range control delivers enhanced robustness versus target- only approaches.
- Easy maintenance – Range control design enables easier tuning and enhanced performance: robust control design reduces tuning needs.

Predict-Back Feature on Furnace Controller

The predict-back technology provides a more advanced feature of Honeywell Profit Controller. Predict-back provides additional feed forward information and disturbance rejection to Controlled Variables (CV) when the process value (PV) of a Manipulated Variable (MV) or a Disturbance Variable (DV) deviates from its setpoint (SP). Predict-back is particularly useful when the disturbance rejection of the MV or the DV is slower than the effects of the unrejected (leak-through) disturbances on the CVs.

When an MV is the SP of DCS-based PID controller, it is usually a good practice to keep the PID cascade closed if the control loop performs well, and use their SPs as MVs within a Profit Controller design. Keeping higher-level cascades closed can help linearize the process, improve disturbance rejection and thus improve Profit Controller performance across different operating conditions. Examples include heater outlet temperature, tower overhead temperature, accumulator pressure, etc.

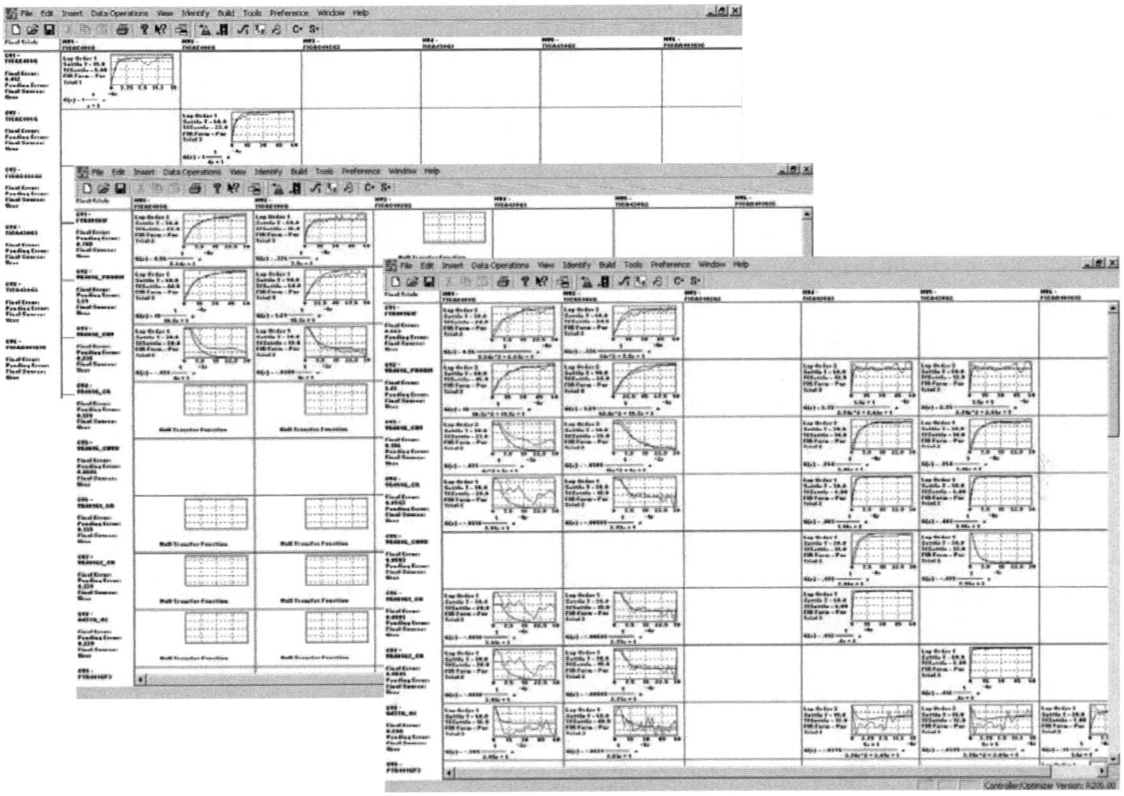

Fig 2 - Vapor Phase Furnace Model with Predict-back Model and Estimated Disturbance Model

Honeywell applies Predict-back on the Furnace Profit Controller. The dynamic responses of higher level PID (i.e. preheat temperature controller, reboiler steam controllers, coil outlet temperatures) cascades is generally slower than the basic regulatory controller (i.e. steam flow controllers, fuel pressure controllers), thus disturbances leak through these MVs can take longer to reject. In a traditional MPC controller design where only the SP of the MV is considered, process disturbances that result in a deviation of the MV.PV from its SP are not detected until the CVs are affected. When the CVs do respond and violate their limits, Profit Controller will move the MVs accordingly, even through the MV.PV may have already returned to its SP.

Product Value Optimization (PVO)

An Excel-based PVO analysis tool is used by engineering personnel to understand the optimization moves that the application is making. The tool uses the current gain information, and pricing information from plant and then calculates the equivalent LP costs for each MV as well as the decomposition matrix represented as MV priorities. The spreadsheet provides insight to assist the engineer in understanding the moves that Profit Controller is making. In addition, the tool is useful in predicting and validating the optimization direction.

The objective function in general can be linear, quadratic, or nonlinear. For most applications, a steady-state quadratic objective (2) is versatile enough, which is the default objective function form in Profit Controller.

$$J = c_x{}^t\, x_{ss} + c_y{}^t\, y_{ss} + (x_{ss} - x_o)^t \mathbf{D}_x (x_{ss} - x_o) + (y_{ss} - y_o)^t \mathbf{D}_y (y_{ss} - y_o)$$

subject to $CV_L \leq y_{ss} \leq CV_H$

$MV_L \leq x_{ss} \leq MV_H$

where

- J is the unit optimization cost objective function
- c_x is MV cost coefficient
- c_y is CV cost coefficient
- x_{ss} is MV steady state value
- y_{ss} is CV steady state value
- x_o the MV desired value
- y_o the CV desired value
- Dx is the cost of deviating from MV desired value
- Dy is the cost of deviating from CV desired value

	A	B	C	D	E	F	G	H	I	J	K	L	M	N	O
1				**MV Name**	FICA402C	FICA402D	FICA402E	FICA402F	TICA430C	TICA430D	TICA430E	TICA430F.SP			
2				**Equiv. LPW**	-14.2842	-14.2842	-14.2842	-14.2842	-1721.17	-1721.17	-1721.17	-1721.17	0	0	0
3	**CV Name**	**Min.**	**Max.**	**LP Weight**	**1694.78**	**1694.78**	**1694.78**	**1694.78**	0	0	0	0			
4	VCM_PROD.PV	1	1	**-462**	*3.823*	*3.823*	*3.823*	*3.823*	3.764	3.764	3.764	3.764			
5	VR401_CNV.PV	1	1	0	-0.050	-0.050	-0.050	-0.050	0.193	0.193	0.193	0.193			
6	VR401C_CR.PV	0	1	0	-0.132	0	0	0	0.70	0	0	0			
7	VR401D_CR.PV	0	1	0	0	-0.132	0	0	0	0.70	0	0			
8	VR401E_CR.PV	0	1	0	0	0	-0.132	0	0	0	0.70	0			
9	VR401F_CR.PV	0	1	0	0	0	0	-0.132	0	0	0	0.70			
10	PT405C.PV	0	1	0	2.489	0	0	0	0.463	0	0	0			
11	PT405D.PV	0	1	0	0	2.489	0	0	0	0.463	0	0			
12	PT405E.PV	0	1	0	0	0	2.489	0	0	0	0.463	0			
13	PT405F.PV	0	1	0	0	0	0	2.489	0	0	0	0.463			
14	VR401C_CNVD.PV	1	1	0	-0.165	0.055	0.055	0.055	0.337	-0.112	-0.112	-0.112			
15	VR401D_CNVD.PV	1	1	0	0.055	-0.165	0.055	0.055	-0.112	0.337	-0.112	-0.112			
16	VR401E_CNVD.PV	1	1	0	0.055	0.055	-0.165	0.055	-0.112	-0.112	0.337	-0.112			
17	VR401F_CNVD.PV	1	1	0	0.055	0.055	0.055	-0.165	-0.112	-0.112	-0.112	0.337			
18	VR401C_TMT.PV	0	1	0	1.586	0	0	0	0.832	0	0	0			
19	VR401D_TMT.PV	0	1	0	0	1.586	0	0	0	0.832	0	0			
20	VR401E_TMT.PV	0	1	0	0	0	1.586	0	0	0	0.832	0			
21	VR401F_TMT.PV	0	1	0	0	0	0	1.586	0	0	0	0.832			
22	PICA455C.PV	1	1	0	0.345	0	0	0	0.121	0	0	0			
23	PICA455E.PV	1	1	0	0	0	0.345	0	0	0	0.121	0			
24	PICA455F.PV	1	1	0	0	0	0	0.345	0	0	0	0.121			
25	PICA458C.PV	1	1	0	0.345	0	0	0	0.121	0	0	0			
26	PICA478D.PV	1	1	0	0	0.345	0	0	0	0.121	0	0			
27	PICA458E.PV	1	1	0	0	0	0.345	0	0	0	0.121	0			
28	PICA458F.PV	1	1	0	0	0	0	0.345	0	0	0	0.12099			
29	FICA458C.PV	0	1	**6.48**	8.799	0	0	0	2.753	0	0	0			
30	FICA478D.PV	0	1	**6.48**	0	8.799	0	0	0	2.753	0	0			
31	FICA458E.PV	0	1	**6.48**	0	0	8.799	0	0	0	2.753	0			
32	FICA458F.PV	0	1	**6.48**	0	0	0	8.799	0	0	0	2.753			
33	FICA402C.OP	1	1	0	0.336	0	0	0	0	0	0	0			

Instructions | GAIN MATRIX | Max CV lim. Cost over -CV | Min CV lim. Cost over CV | Max CV lim.

Fig 3- PVO Structure of Profit Controller Gain Matrix (Furnace Control)

9. Description of Models for EDC Cracking

EDC CRACK is Technip's rigorous yield prediction program for the pyrolysis of di-chloro ethane. It is based upon a mechanistic kinetic scheme of more than 300 reactions, representing the radical chain reaction mechanism of 37 molecular species and 20 radical species. The fundamental basis of the EDC CRACK program allows accurate prediction of yield patterns from H2 through C6 and for furnace parameters such as tube skin temperature, coke build-up, heat flux, coil pressure drop and coil outlet temperature. Given the basic geometry of the pyrolysis reactor, EDC CRACK can simulate the performance of nearly any type of pyrolysis coil in commercial use today. In conjunction with Honeywell, Technip has developed a version of EDC CRACK for use in advanced process control of EDC cracking furnaces ("EDC APC"). EDC APC is the extension to EDC CRACK that allows the rigorous free-radical reactor model to be called directly from a third-party advanced process controller. EDC APC is supplied in the form of a Microsoft Windows DLL. The model is accessed via a FORTRAN subroutine call. The furnace geometries are stored in an external binary file that can be configured by the user. For details please refer to Wagner et. al [5].

10. Detailed description of an enabling technologies

In addition to MPC and VCM yield models, the following enabling technologies are needed for VCM APC applications.

Honeywell VCM Profit Toolkit

Honeywell VCM Profit Toolkit is an ensemble of generic programs that have been specifically developed to enhance Profit Suite applications for VCM plants.

Operation of EDC cracking furnaces is inherently non-linear due to complex chemical reactions in the crackers. It is desirable to keep the furnace operating for a fairly long period of time (up to a year in some cases). As furnace operation continues, coke (a byproduct of cracking reactions) deposits on the inside of the cracking tubes furnaces that increases resistance to heat transfer and pressure drop in the coil, this also changes the EDC conversion, VCM yield and coking rate characteristics further adding to operational non-linearities. EDC conversion, VCM yield and coking rates are important (controlled) variables in VCM plant operation, there usually is no direct measurement of these variables. Using EDCAPC model from Technip, VCM Profit Toolkit can provide an accurate estimation of these variables (CVs in the controller) and thus provide a continuous feedback to Profit Controller. In addition, VCM Profit Toolkit maintains coke thickness profile in each coil, which is an unmeasured quantity. This in essence is the functionality of the VCM Profit Toolkit program.

VCM Profit Toolkit suite comprises software components: (1) to collect process data (2) to schedules execution of VCM Profit Toolkit application which perform cracker simulation (3) to deposit results from VCM Profit Toolkit to process (4) to simulate current furnace operation using Technip's EDCAPC module which has been configured for a particular furnace.

VCM Profit Toolkit can be executed in one of the following 3 modes; (1) **ON mode**: For on-line closed loop usage. This mode is configured as a toolkit application that contains three toolkits, the first toolkit collects process data and stores them in the database file, the second toolkit initiates execution command by using process from this database. EDCAPC is called to perform simulation for all furnace coils that calculates furnace CVs and stores them in the database file. The third toolkit accesses the database file and deposits the predicted CV values to the process. Execution in this mode is automatic at a frequency set in the toolkit application. (2) **WRM mode**: For on-line open loop usage. WRM mode is identical to ON mode, except that the results are sent back to the process. (3) **OFF mode**: OFF mode is similar in functionality to other modes except that the process data is static and comes from the offline directory and simulation results are not sent to process.

Honeywell Uniformance (Process Historian Database)

It is an advanced plant information management system, which collects, stores and replays critical plant data. PHD and Profit Controller online software were installed on the APC computer. The software that provides a graphical user interface to the Profit Controller, Profit Viewer, is installed on operator stations. Once APC computer is connected to the control network, Profit Controller is ready to be turned on.

11. Results from actual projects [2, 3 and 4]

Cracking Furnaces:

In general, the control of furnace feed flows and coil outlet temperatures improve significantly after the implementation of APC as furnace inlet conditions can be set more precisely to maintain objectives. VCM production must be adjusted to meet upstream and downstream requirements, there is great incentive to provide a control solution that improve VCM production and improve the stability of the units operation.

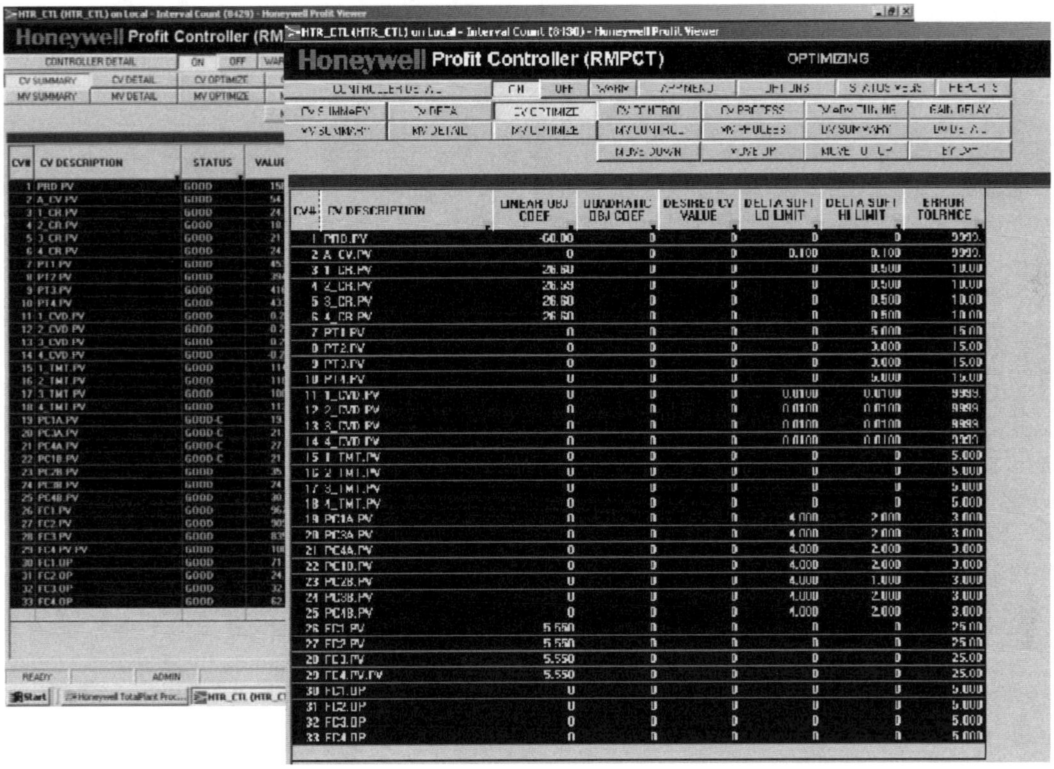

Fig 4 – Profit Viewer of Cracking Furnaces Control

Profit Controllers in conjunction with the EDCAPC model, enables a VCM producer to raise the EDC conversion (or cracking depth) without the fear of over-cracking in any of the furnace passes. This results in less EDC recycle and hence reduced fuel gas consumption per pound of VCM produced. At times when demand allows peak production the reduced recycle means increased furnace capacity and higher peak production rates.

Chart 1 shows reduced steam usage per pound of VCM produced and Chart 2 shows the reduced furnace fuel gas consumption per pound of VCM produced.

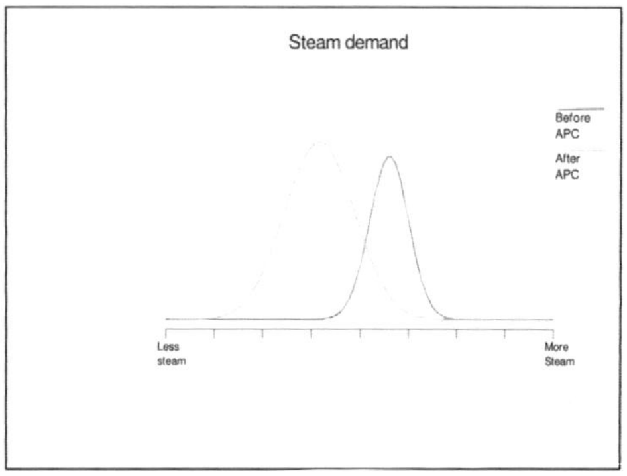

Chart 1 - Fuel Ratio

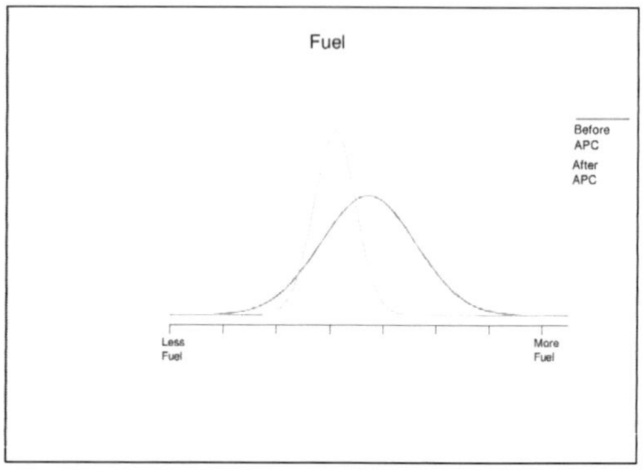

Chart 2 - Steam Ratio

Charts 3 and 4 also provide similar results showing the reduced utilities.

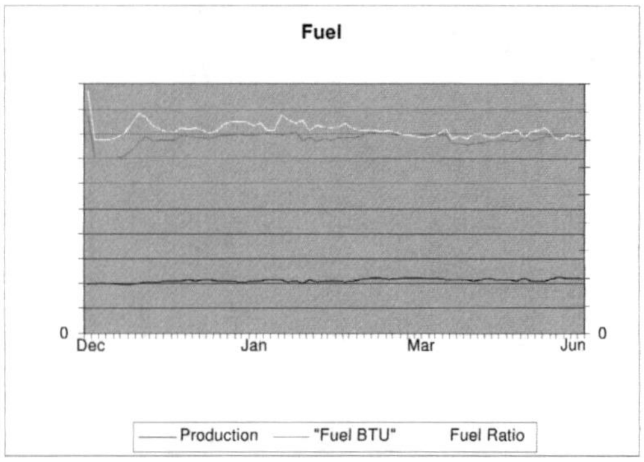

Chart 3 – Production, Steam and Steam Ratio

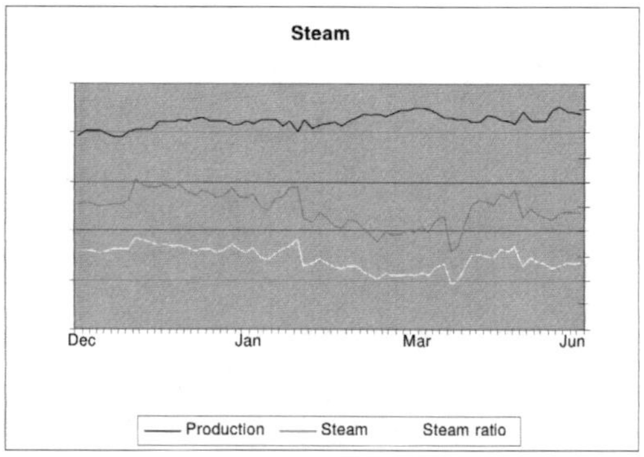

Chart 4 – Production, Fuel and Fuel Ratio

Furnace pass balancing:

Figure below shows the furnace coil outlet (COT) temperature difference in each of two passes before and after the Profit Controller.

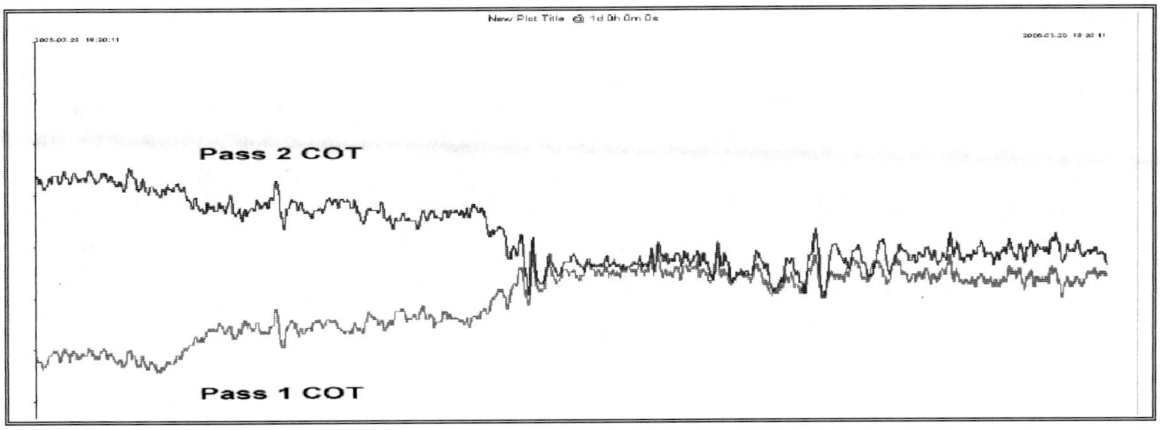

Fig 5 – Furnace pass balancing

HCl columns:

In the HCl columns, improved stability helps reduce steam usage while maintaining recovery of HCl. The major sources of HCl Column upsets are (1) the changing composition of the feeds due to variations in cracking depth and (2) for plants with two HCl columns, the variations in feed allocation between the two columns. (3) Very high non-linearity of the process at the top zone of the HCl Column. Implementation of Profit Controller on the Cracking Furnaces in turn reduces the first source of variation. The second source is reduced by including feed flow allocation in the control strategy to equally distribute the feed into the HCl Columns, after which the Profit Controller on HCl Columns is able to control the top zone temperature using temperature profile control to split HCl from VCM.

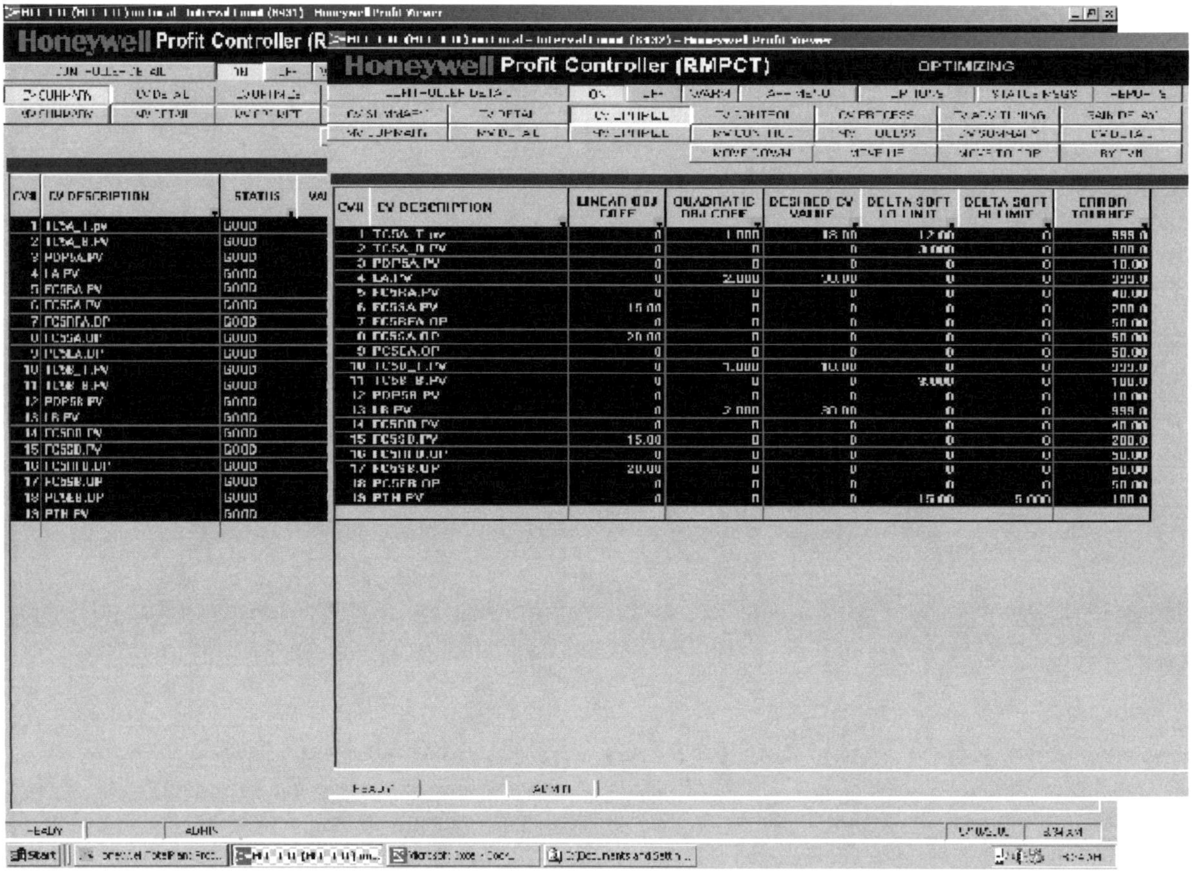

Fig 6 – Profit Viewer of HCl Column Control

Chart 5 shows a reduction in steam usage at balanced feed flows to the HCl columns. Chart 6 shows a reduction in column reflux flows at balanced feed flows to HCl Columns.

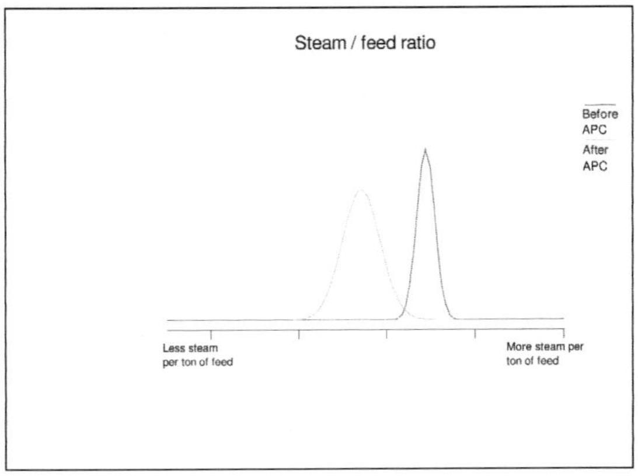

Chart 5 – Steam/Feed Ratio

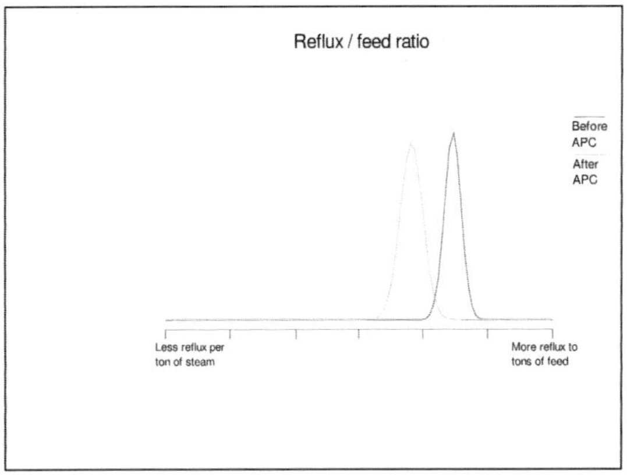

Chart 6 - Reflux/Feed Ratio

Production maximization:

Figure below shows that the furnace feed pass flow was maximized against the constraints from HCL column pressure, EDC vaporizer pressure and COT difference.

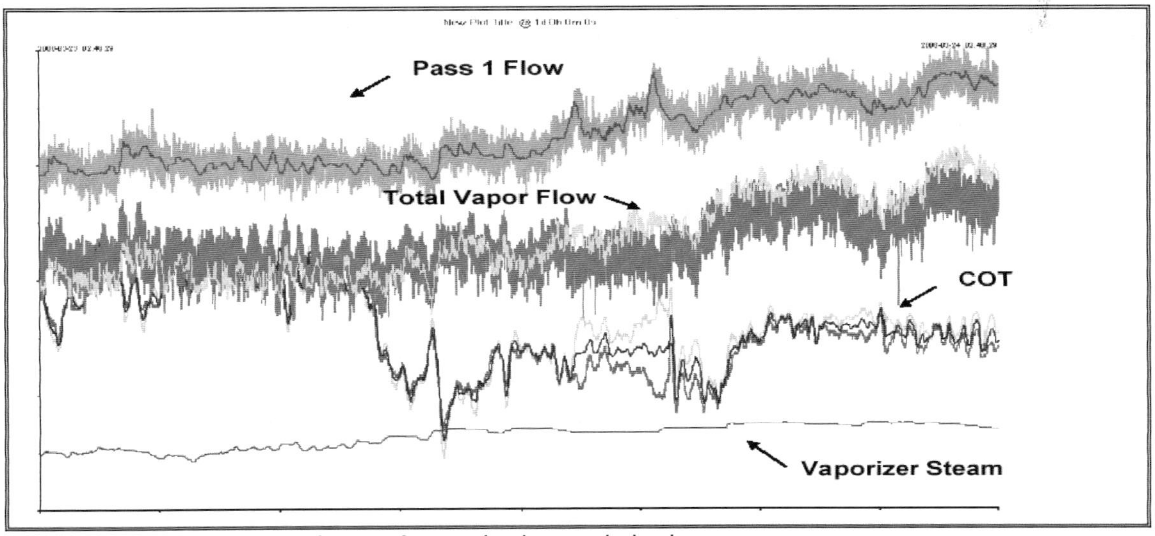

Fig 7 –VCM production maximization.

Feed maximization:

Figure below shows that the furnace the total feed rate flow was maximized against the constraints from HCL column top pressure.

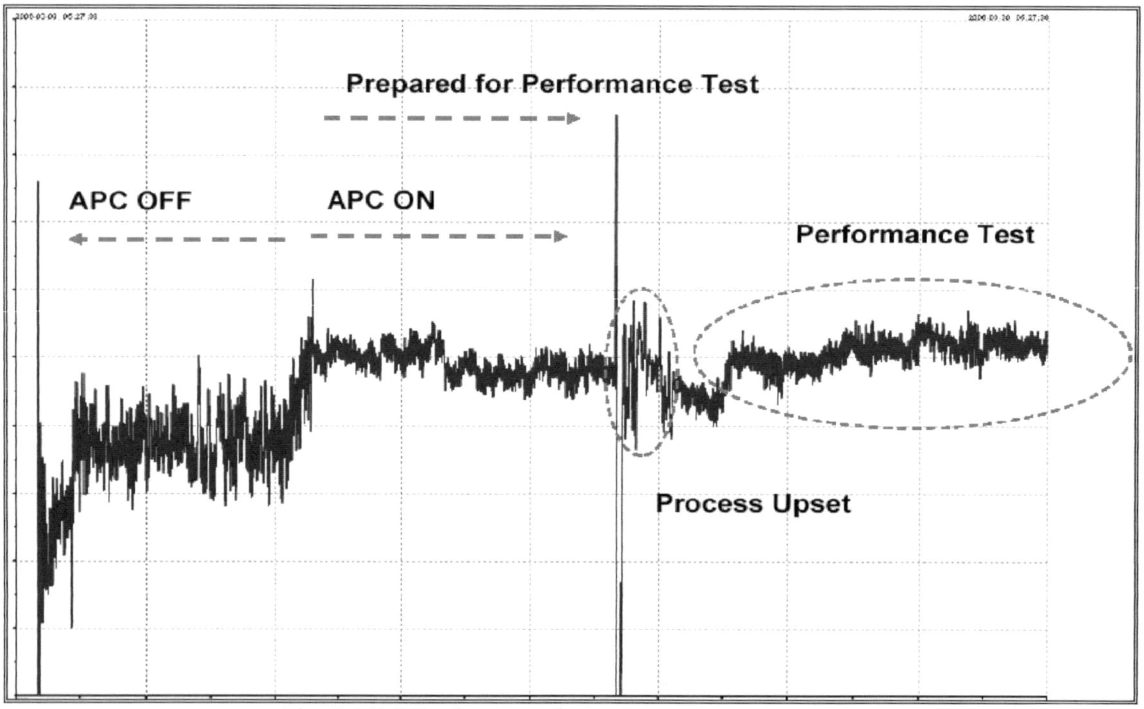

Fig. 8: EDC Feed rate maximization

Benefits Summary

Benefits of APC for a VCM plant are very significant, typical payback is usually between three to six months.

As the figures and charts above show, the Advanced Control on VCM plant reduces raw material and energy usage per unit of production. Alternatively, higher production can be achieved for same level of raw material and energy usage.

Keys to Project Success

Key to the project success is operator acceptance of APC, their involvement must be sought right from the beginning of the project. Another factor for success is strong management support that allows plant control engineers to fully participate and take responsibility in all project phases including controller design, basic controller tuning, plant step testing, implementation and commissioning of the multivariable predictive controllers. Rigorous attention and resolution of regulatory control problems before implementation of the advanced controls is considered a major factor in achieving a high level of robust control.

12. Conclusion

Advanced Process Control of VCM plants is particularly challenging since many of the key variables in the cracking furnace such as EDC conversion and VCM production are not usually not measured. Use of high fidelity EDC cracking model used as a sensor makes the application of Multivariable Predictive Controllers possible. Application of such APC methodology to VCM is very beneficial, such projects typically payback well within a year.

Acknowledgements

The authors would like to thank the Honeywell and Technip management for permission to publish this paper.

References

[1] Anonymous, "*PROFIT Optimizer (DQP) Concepts and Implementation Course Notes*", Honeywell Training Document (1998).

[2] DiBiano, R. et. al. "*Real-time Modeling & Multivariable Control of a Vinyl Chloride Monomer Cracking Furnaces*", paper presented at the Honeywell User's Group Conference, Phoenix, AZ 2004.

[3] Lee, B. et. al. "*Dynamic Optimization of Profit and Olefin Production using Profit Controller and Profit Optimizer in Ethylene Plant*", paper presented at the Honeywell User's Group Conference, Phoenix, AZ 2004.

[4] Novak, M. et. al. "*Real-time Modeling & Multivariable Predictive Control of Vinyl Chloride Monomer Cracking Furnaces and HCl Columns*", paper presented at the Honeywell User's Group Conference, Phoenix, AZ 2006.

[5] Wagner, E. et. al., "*A new simulation program predicts EDC furnace performances*", La Chimica & L'Industria, Vol 75, 1993.

Direct Control of Chemical Processes with Multi-Rate Measurements

Sairam Valluri and John Jordan. Control Systems Engineering, Lyondell Chemical Company, 2502 Sheldon Road, PO Box 30, Channelview, TX 77530

This paper presents a methodology for direct control of industrial chemical processes with measurements coming to the process control system at different rates. Many measurements are fast and available to the closed-loop controller almost instantaneously. However, some measurements are slow and pose challenges in closing the controller loop along with the fast measurements. Indirect control of fast measurements that help control the slow measurements within certain bounds is one traditional approach. However, indirect control may result in significant variations in the slow measurements that are often key process and economic indicators. On the other hand, direct model-based control of slow measurements can result in tighter and consistent operations that render significant improvements in production rates and process economics. Model-based controller implementations and sustained controller performance needs for an industrial chemical production process are illustrated.

Model Based Online Analysis and Monitoring of Tank Qualities

Suresh S. Agrawal, Offsite Management Systems LLC, 3311 Stoney Mist Dr., Sugar Land, TX 77479

A typical refinery or chemical plant has a tank farm consisting of 50-300+ tanks used to receive, store and dispatch material in and out of it. The manufacturing process dictates that the amount and quality of tanks contents be known to all concerned plant departments as soon as possible for their business decisions.

In an automated plant, the tank inventory is obtained online by using an Automatic Tank Gauging (ATG) system whereas the tank quality information relies only on manual samples and their lab analysis. Typically, a plant spends 2-8 MUS$/year on lab analysis of tank contents. The quality information of tank contents is not possible in near real-time even if there are analyzers at the tank inlet or outlet as they indicate qualities only of inlet / outlet streams and not of tanks while loading or unloading a tank. Consequently, any decision or action by refinery planners and operations management, based on tank quality, may be outdated soon if the tank has been active during the decision process.

This paper discusses successful implementation of a sophisticated and state-of-the-art Tanks Quality Tracking System to provide open-loop online tracking qualities of run-down tanks of three Diesel and two Kero blenders and gasoline blender components tanks in a refinery in Asia-pacific region. The system has helped the refinery to reduce their qualities giveaways of 90%, Sulphur, flash and freeze points to acceptable limits and minimize reblends of off spec products. The system has been fully operational at the refinery since August, 2002 and has flexibility, robustness and features to incorporate any of the built-in 100+ linear / non-linear models for the qualities of crude, gasoline, diesel, fuel and other products' tanks considering all possible configuration of inlet / outlet flows, lab analysis of tank samples, inlet / outlet analyzers and other factors. The system has client-server architecture resident on Windows NT/2000 platform with SQL Server /Oracle DBMS and interfaces with any DCS system and field signals. Furthermore, ROI of such a system is less than one month based on the estimate of realizable tangible benefits and was customized and implemented within two months. The system is also plug N' Play upgradeable with either already installed or third party's feedback control or optimizer module.

Development of in-Situ Analysis for the Chemical Industry

J.D. Tate, The Dow Chemical Company, 2301 N. Brazosport Blvd., Freeport, TX 77541 and Trevor Knittel, Analytical Specialties Inc., 910 Gemini, Houston, TX 77058.

The goal of this project was to develop two platforms for robust, in- or at-line, process measurements that provide data at/or near real time. Technologies targeted were Laser Based Sensors and Solid State Sensors. These systems are capable of operating at the required process conditions (pressure, temperature, etc.), providing real time or near real time data, and significantly reducing installation and operational cost. The need for these platforms, and the measurements they will provide, are driven by advances in process control (APC) systems and the need to "close the loop" in modern control systems. This program is complete and has resulted in successful development and commercialization of two laser based analyzers. Program information, application and resulting products/installations will be presented.

Process Monitoring Using Robust Chemometric Spectrum Models for Predicting Concentration Profiles

Uwe Kruger[1], Yan Zhou[2], David Rooney[3], Xun Wang[2], and Jillian Thompson[3]. (1) Electronics, Electrical Engineering and Computer Science, Queen's University Belfast, Ashby Building, Stranmillis Road, Belfast, BT9 5AH, United Kingdom, (2) Electronics, Electrical Engineering and Computer Science, Queen's Univeristy Belfast, Ashby Building, Stranmillis Road, Belfast, BT9 5AH, United Kingdom, (3) Chemistry and Chemical Engineering, Queen's University Belfast, David Keir Building, Stranmillis Road, Belfast, BT9 5AG, United Kingdom

Heterogeneous catalytic processes are of extreme importance to the entire range of chemical industries, as the reactions that are being carried out increase in complexity, so to must the tools used to analyze and interpret the reaction data. Robust and accurate chemometrics/spectrocopy methods allow rapidly determination of the concentrations within reactors operating under industrial reaction conditions thereby enabling on-line optimization of such reaction conditions to maximize the desired product and reduce both energy costs and waste. The papers listed below discuss some of the techniques used for in situ monitoring on bioprocess, which are particularly sensitive to reaction environments and therefore usually have tighter controls.

This is the reason why research in chemical kinetics in addition to the measurement of reaction rates also involves the study of reaction intermediates by computation, spectroscopy, isotopic labeling, etc. Once reliable data for key intermediates become available these data constraints the reaction mechanisms and thereby the predicted macroscopic kinetics. This approach enables the discrimination between different reaction mechanisms. The availability of a robust PLS model provides a technique by which an estimation of the local concentrations from spectroscopic data, i.e. Raman, IR or ATR intensities, can be obtained at a very short timescale. The obtained concentration estimates of each reacting species can be subsequently used to recursively update the reaction rate coefficients in a given model, which typically change with temperature for example.

For more complex systems such as those often encountered in heterogeneously catalyzed reactions, the performance of the catalyst may also change with temperature or the reaction may switch to a different mechanism and hence, the structure of the defining rate equations can also vary. Using a total of three sets of experimental data, a Raman spectroscopy and a calibration experiment as well as recorded data from a debutanizer unit, the contribution of the presented work are two detailed application study which involve (i) recorded data from a Raman spectroscopy experiment and (ii) data from a calibration experiement. These applications highlight the merits and requirements of a robust PLS parameter estimation and show the practical usefulness of newly developed diagnosis charts to determine which Type a specific outlier is assoicated with. The presentation finally shows how to utilize these PLS models to identify abnormal process behavior through the on-line analysis of concentration estimates.

References

Lynggaard, H., Andreasen, A., Stegelmann, C., Stoltze, P., "Analysis of simple kinetic models in heterogeneous catalysis", Progress in Surface Science, Vol. 77, No. 3-4, pp. 71-137, 2004.

Vojinovi•, V., Cabral, J.M.S., Fonseca, L.P., "Real-time bioprocess monitoring Part I: In situ sensors", Sensors and Actuators B: Chemical, Vol. 114, No. 2, pp. 1083-1091, 2006.

AUTHOR INDEX

American Institute of Chemical Engineers
3 Park Avenue
New York, NY 10016-5991

ISBN: 978-1-60423-363-6